PREFACE

Anyone involved with industrial decisions in today's world must certainly be aware of the need to produce products which are more productive and more cost efficient compared to even a few years ago. Increased productivity and cost efficiency are goals which often compete with one another while government regulations can make the task even more difficult.

Better hydraulic systems are required to produce those more efficient machines. Better hydraulic systems require better components and fluids. Human comfort requires noise control and user friendly interfaces. Fluids must work at higher temperatures, pressures, and meet environmental regulations.

This SAE special publication, <u>Fluid Power: Applications, Standards, Noise, Lubricants, and Testing</u> (SP-1110), combines papers which advance fluid power technology in way that can be used and understood by applications engineers and designers. Papers have been drawn from both industry and academia. The highly theoretical information is not included in order to focus on the applications. What is presented is useful information to help push the technology envelope. Fluid power definitely has its place in the future.

Kenneth Stoss
Chairman, Fluid Power Committee

Fluid Power: Applications, Standards, Noise, Lubricants, and Testing

SP-1110

GLOBAL MOBILITY DATABASE

All SAE papers, standards, and selected books are abstracted and indexed in the Global Mobility Database.

Published by:
Society of Automotive Engineers, Inc.

ISBN 1-56091-688-5
SAE/SP-95/1110
Library of Congress Catalog Card Number: 95-70542
Copyright 1995 Society of Automotive Engineers, Inc.

Persons wishing to submit papers to be considered for presentation or publication through SAE should send the manuscript or a 300 word abstract of a proposed manuscript to: Secretary, Engineering Meetings Board, SAE.

Printed in USA

TABLE OF CONTENTS

952075

Performance of Selected Vegetable Oils in ASTM Hydraulic Tests

Lou Honary
University of Northern Iowa

ABSTRACT

The initial goal of this project was to investigate the potential uses of soybean oil as an industrial hydraulic fluid. Previous test results had indicated that the concern with vegetable oil was mainly with the lack of inherent oxidative stability. In general all of the vegetable oils that had been tested, had performed well in terms of wear protection in ASTM D-2281 (100-hour) tests. The project activities included analysis of several vegetable oils in ASTM D-2271 (1000-hour) wear tests, using the 104-C Vickers vane pump. The longer test period was selected to observe changes in the viscosity [as a measure of oxidative stability] of the test oils. The results indicated that the fatty acid profiles of vegetable oils play a significant role in their oxidative stability in industrial lubricant applications. The research results should help to identify the needed genetic modification of oil seeds to produce oils more suitable for industrial applications.

INTRODUCTION

As a follow up to an earlier paper, Honary (1994), the performance of several vegetable oils in ASTM D-2271 are reported here. The primary goal of this project was to investigate the potential use of soybean oil as an industrial hydraulic fluid. The report, therefore, provides information on the performance of soybean oils with and without additive in various tests. Tests results include information on changes in viscosity (using ASTM D-445, changes in Total Acid Number (TAN) using

ASTM D-664, and changes in elemental composition (ppm) using Inductive Coupled Plasma test.

The ASTM D-2271 tests require a Vickers 104-C vane pump to be operated at 1200 rpm for 1000 hours @ 6.89 +/- .14 MPa (1000 +/- 20 psig) and @ 79°+/- 3°C.

The Vickers 104-C vane pump cartridge includes four major components in addition to the 12 vane pieces. The performance of the oil is judged by the onset of metal wear as measured on the pump cartridge. Figure 1 shows the test set up as per ASTM.

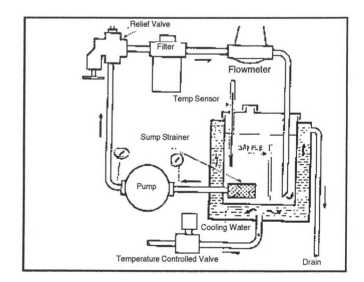

Figure 1: ASTM Pump Wear-Test Set-up as presented by ASTM (1986)

Oil Type			Percentages of Fatty Acid Compositions						
			16:0	16:1	18:0	18:1	18:2	18:3	Δ Viscosity in cSt. 1000-hour
Normal Soy W/O Additives	Oil 1		11	0.1	4.2	23.4	52.5	8.8	43.86
Low Linolenic Soy W/O Additives	Oil 2		9.7	0.1	4.2	32.9	50.3	2.9	39.56
Choice Soy Oil W/O Adtvs. (c:14=.07)	Oil 4		9.5	n/a	4	37.5	42.31	4.75	24.18
HOSO W/O Additives	Oil 6		3.5	0.1	4.8	78.2	13.3	0.1	19.24
HOCO W/O Additives	Oil 7		3.4	0.2	2	76.5	8.1	7	19.53
UHOSO W/O Additives	Oil 8		4	0	1.4	86.8	6.2	0.1	16.23
Normal Soy W. Additives *	Oil 3		11	0.1	4.2	23.4	52.5	8.8	20.75
Low Linolenic Soy W. Additives	Oil 5		9.7	0.1	4.2	32.9	50.3	2.9	20.14
Choice Soy Oil W. Adtvs. (c:14=.07)	Oil 9		9.5	n/a	4	37.5	42.31	4.75	7.84

* Modified test used a Vickers 20-VQ pump

Table 1: Relationships between fatty acid composition of various vegetable oils and changes in their viscosity in ASTM D-2271

DISCUSSION

Table 1 shows the results of several 1000-hour tests using various vegetable oils. An analysis of the changes in the viscosity of these oils indicates that [normal] crude soybean oil shows considerably higher (increased) viscosity than oils that are genetically modified to alter their fatty acid. Ultra high oleic sunflower oil, for example, with an oleic value of 86.6 showed considerably lower changes in viscosity than all the other oils.

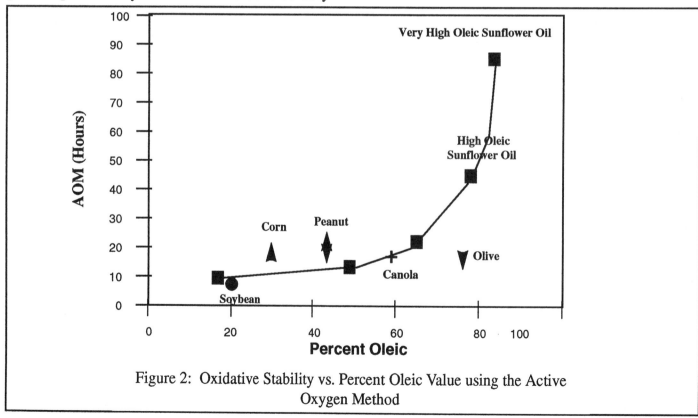

Figure 2: Oxidative Stability vs. Percent Oleic Value using the Active Oxygen Method

Note: From Naegley, P.C. (1992) Environmentally Acceptable Lubricants. Lubrizol Corporation, Wicklife, Ohio.

2

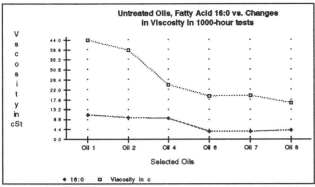

Figure 1: Relationships between Palmitic Acid and Viscosity Change in ASTM D-2271 (1000-Hour) Tests

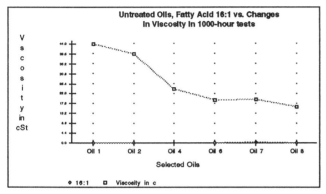

Figure 2: Relationships between Palmitoleic Acid and Viscosity Change in ASTM D-2271 (1000-Hour) Tests

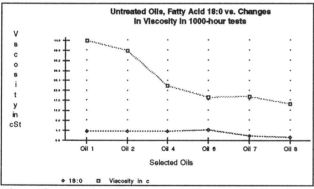

Figure 3: Relationships between Stearic Acid and Viscosity Change in ASTM D-2271 (1000-Hour) Tests

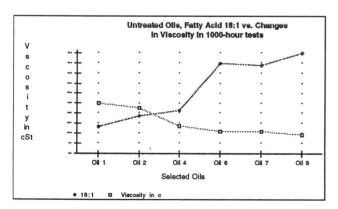

Figure 4: Relationships between Oleic Acid and Viscosity Change in ASTM D-2271 (1000-Hour) Tests

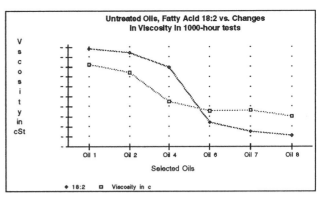

Figure 5: Relationships between Linoleic Acid and Viscosity Change in ASTM D-2271 (1000-Hour) Tests

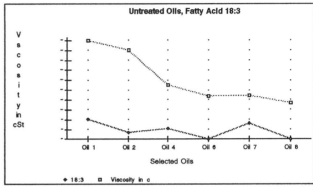

Figure 6: Relationships between Linolenic Acid and Viscosity Change in ASTM D-2271 (1000-Hour) Tests

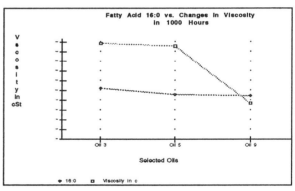

Figure 7: Relationships between Palmitic Acid and Viscosity Change in ASTM D-2271 (1000-Hour) Tests

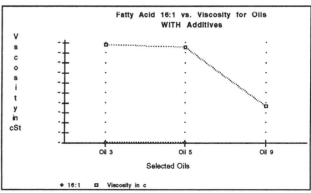

Figure 8: Relationships between Palmitoleic Acid and Viscosity Change in ASTM D-2271 (1000-Hour) Tests

3

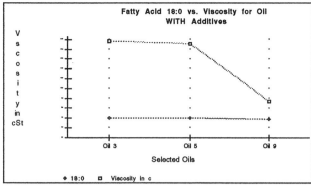

Figure 9: Relationships between Stearic Acid and Viscosity Change in ASTM D-2271 (1000-Hour) Tests

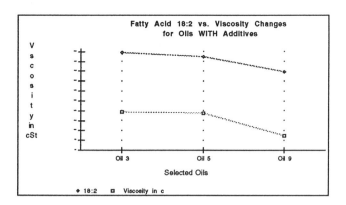

Figure 11: Relationships between Linoleic Acid and Viscosity Change in ASTM D-2271 (1000-Hour) Tests

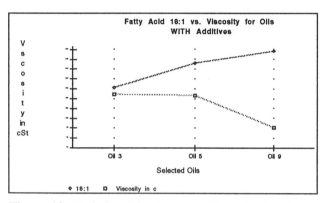

Figure 10: Relationships between Oleic Acid and Viscosity Change in ASTM D-2271 (1000-Hour) Test

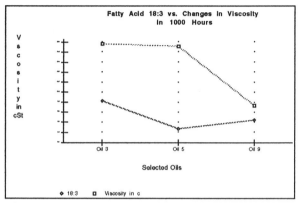

Figure 12: Relationships between Linolenic Acid and Viscosity Change in ASTM D-2271 (1000-Hour) Tests

When only the oleic acid is considered, the results, presented in Table 1, correspond to those determined by Naegley (1992), as shown in Figure 2. Naegley used the Active Oxygen Method in which air is blown through a sample at 98°C and the time to reach a peroxide value of 100 is measured. Table 1, however, shows the percentages of other fatty acids which also play a role in the overall performance of the vegetable oils in such industrial applications as hydraulic oil. Figures 2 through 13 show relationships between individual fatty acids and changes in viscosity. Three of the oils were mixed with additive packages and then tested, as shown in Figures 7 through 12. Each point on the chart represents the viscosity change in a 1000-hour test. In both cases the results indicated that the percentage of oleic fatty acid directly correlates with the improved oxidative stability and thus lesser change in the oil viscosity in 1000-hour of pump testing.

Figure 5, shows the relationships between linoleic acid and changes in viscosity. In this case, the data indicated that lower linoleic values result in the improved oxidative stability and thus lesser change in the oil viscosity in 1000-hour of pump testing. Similar results were observed for the linolenic acid as shown in Figure 6.

The use of additive packages did have an impact on the overall oxidative stabilities of the test oils. The impact of the fatty acids of those oils appeared to remain the same in oils with additives as in the oils with out additives.

The choice soybean oil with its fatty acid profile as presented in Table 1, was a chemically modified (partially hydrogenated) oil designed for improved oxidative stability. The results indicated that when combined with the proper additive package, the oil showed desirable oxidative stability. The choice oil when tested in the ASTM D-2271, resulted in 40 mg pump wear. Combined with 7.84 cSt change (14.7%) in viscosity, this oil has shown very good potential for use as a hydraulic fluid. The same oil was tested in piston pumps at one of the John Deere plants for 1000 hours with minimal change in viscosity (1 cSt) and passed visual wear protection requirements. Further field testing of that oil is underway and the results will be published in future papers.

Conclusions

The data indicated that fatty acid profiles of vegetable oils have significant impacts on their oxidative stability. It was further observed that changing the fatty acid profile through genetic modification of the oil seed or through chemical modification (partial hydrogenation) of the oil affects the oxidative stability of the oil. The results further suggest that, the often food related, research on genetic modification of the oil seeds would have potential for creating seed oils suitable for industrial applications.

References

Honary L. A. T. (1994). Potential utilization of soybean oil as industrial hydraulic oil. SAE Technical Paper # 941760. Warrandle, PA: SAE Publications.

Honary L. A. T. (1994). "Soybean oil as an alternative hydraulic fluid". Fluid Power Journal. Milwaukee, WI: Fluid Power Society.

Naegley, P. C. (1992). Environmentally acceptable lubricants. Lubrizol Corporation, Wicklife, Ohio.

Historical Overview of the Development of Water-Glycol Hydraulic Fluids

G. E. Totten and R. J. Bishop, Jr.
Union Carbide Corp.

ABSTRACT

Because of on-board fire problems during World War II, the us Navy initiated a program to develop hydraulic fluids that were more fire-resistant than the mineral oils that were in use at that time. Water-glycol hydraulic fluids were subsequently developed and first commercialized in 1947 which offered vastly improved fire resistance relative to mineral oils. Since 1947, in addition to formulation changes, there is significantly greater understanding of the impact of these changes on pump wear performance. This paper will present a selected overview of water-glycol formulation chemistry, some of the fluid formulation issues that have been encountered and the evolutionary improvement of hydraulic pump wear performance.

INTRODUCTION

Hydraulic systems use fluids for energy transmission. In 2000 B.C. the ancient Egyptians provided some of the earliest recorded uses of fluid power with their water-driven devices.[1] Hydraulic theories were developed by Pascal (1650) and Bramah (1795). However, industrial hydraulic systems were not developed in large scale until the early 1900's.[1] The most common hydraulic fluids were, and continue to be, derived from mineral oil.

In 1943, during World War II, the navy experienced numerous on-board ship and aircraft disasters which were due to hydraulic line rupture and subsequent ignition of the resulting sprays. This led to the development, by the Naval Research Laboratory, of a new class of fire-resistant hydraulic fluids known as *hydrolubes*.[2,4] Hydrolubes, also known as *water-glycol* hydraulic fluids are aqueous, polymer-thickened water-glycol solutions first patented in 1947 by Roberts and Fife.[3]

Water-glycol (W/G) hydraulic fluids are part of a larger class of fire-resistant hydraulic fluids, which currently include polyol ester, phosphate ester, water-oil emulsions, invert emulsions and high water base hydraulic fluids (HWBHF).

The current ISO designation for W/G hydraulic fluids is "HFC".[7] Therefore, in some references, particularly European, W/G fluids are referred to as HFC fluids.

In this paper, a selected overview of: water-glycol formulation chemistry, some of the fluid formulation issues that have been encountered and the evolutionary improvement of hydraulic pump wear performance is provided.

DISCUSSION

A. W/G Hydraulic Fluid Formulation Chemistry

There are two goals in this section. One goal is to provide a general overview of W/G hydraulic fluid composition. The second goal is to address some of the more critical W/G formulation issues that have been encountered since 1947.

1. Fluid Formulation

The general classes and functions of W/G hydraulic fluid formulation components are summarized in Table 1. In addition to the components shown, other additives may be used as needed such as, dyes for leak detection, antifoam, air release agents, etc.

In addition to the antiwear additives, it is important to recognize that both the water content and corrosion inhibitors that are present in a W/G hydraulic fluid may dramatically affect antiwear performance. In Figure 1, it is shown that increasing water concentration in excess of approximately 40% will result in corresponding decreases in pump wear performance.[16] It has been reported that water concentrations greater than approximately 35% is required to provide the desired fire safety.[6] Therefore, water content should be maintained during use to at least 35-40% and preferably to within +/- 1%.

Especially since W/G fluids contain water, it is important that the corrosion inhibitor be maintained. The corrosion inhibitor chemistry currently used by most W/G hydraulic fluid

Table 1

Components of a "Typical" Water-Glycol Hydraulic Fluid

Component	Purpose
Water	Fire Protection
Glycol	Freeze point reduction and some thickening.
Thickener	To thicken the formulation and to provide adequate film viscosity at the wear contact
Antiwear Additives	To provide mixed film and some boundary lubrication
Corrosion Inhibitors	Vapor and liquid corrosion protection

formulators is a non-nitrite amine containing additive system. The concentration of the corrosion inhibitor, which may be quantitatively measured by "reserve alkalinity" determination, may also affect wear rates as shown in Figure 2.[16] Therefore, the corrosion inhibitor concentration in the W/G hydraulic fluid, which may vary during use, must also be maintained as well.

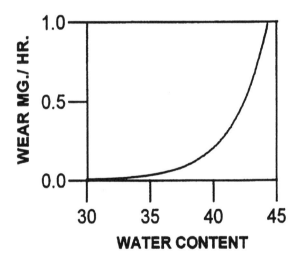

Figure 1 - Effect of water concentration on the pump wear performance of W/G hydraulic fluids.

In addition to maintaining the fluid chemistry during use, it is becoming increasingly recognized that fluid cleanliness must be maintained.[16] Interestingly, in 1947 many industrial hydraulic circuits did not even use filtration! Today, fluid cleanliness levels of ISO 17/14 are recommended. If servo valves are used, ISO 13/10 cleanliness levels are recommended. (The current international standard for cleanliness of a hydraulic fluid is defined by ISO 4406.)

The continued emphasis of fluid cleanliness **and** fluid chemistry improvements have contributed to substantial improvements in the performance of W/G hydraulic fluids since their inception in 1947.

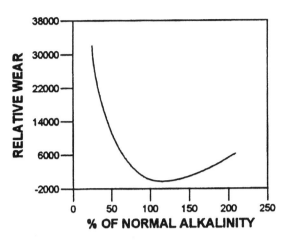

Figure 2 - Effect of corrosion inhibitor concentration on pump wear rates.

2. Shear Stability

Although numerous water soluble polymers have been used as thickeners for W/G fluids, the most common thickener in use in the world today (and in 1947) is poly(alkylene glycol) - PAG. PAG thickeners provide newtonian thickening behavior, which means that the fluid does not undergo shear thinning in the hydraulic pump during use. Also, PAG thickened W/G hydraulic fluids are shear stable which means that the polymeric thickener does not degrade in the high-shear conditions in the hydraulic pump. Shear stability is critically important in view of the very high shear fields, approximately 10^6 sec^{-1}, present in hydraulic pumps.[13]

The newtonian behavior of a recently developed W/G fluid, in shear fields typically found in hydraulic pumps, was recently demonstrated using high-shear capillary viscometry[12]. The results shown in Figure 3, illustrate that a W/G hydraulic fluid containing a "typical" PAG polymer thickener does not undergo any significant shear thinning behavior that would lead to hydraulic leakage which would result in loss of volumetric efficiency during pump operation. Similar high-shear viscometry results have been reported by Isaksson.[15]

A hydraulic fluid may not only undergo viscosity losses when subjected to high shear rates due to non-newtonian behavior, but the polymer used to thicken some non-PAG containing W/G hydraulic fluids may actually undergo molecular degradation (mechano-degradation) resulting in *permanent* viscosity loss. This effect is illustrated in Figure 4 for a series of hydraulic fluids subjected to a 90 minute gear pump test.[14] In this example, it is observed that typical PAG thickened W/G hydraulic fluids were shear stable.

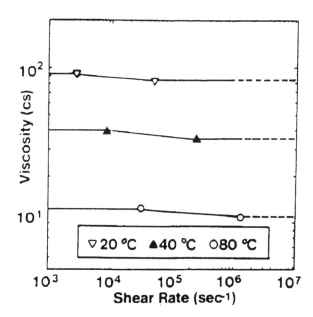

Figure 3 - High-shear viscosity profile of a W/G hydraulic fluid.

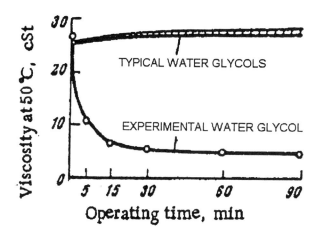

Figure 4 - Gear pump stability test of various W/G hydraulic fluid formulations.

3. Nitrosamine Formation

Some of the earliest corrosion inhibitors used in W/G hydraulic fluid formulation included sodium nitrite and amine nitrites.[8] The technology available at the time required the use of these particular additives in order to achieve the critically necessary solution, vapor and dry-film corrosion inhibition properties. However, recent concerns over potential nitrosamine formation[9] during use of these additives, particularly in metalworking and cosmetic formulation, led to the development of non-nitrite containing W/G hydraulic fluids.[10,11] Nitrite containing additives are no longer used in W/G hydraulic fluid formulations in the USA.

Currently, various amines containing corrosion inhibitor additives are used which provide outstanding corrosion protection in use. The vapor and liquid phase corrosion inhibition properties of a typical W/G hydraulic fluid is illustrated in Figure 5. These inhibitory properties are maintained during use as long as the concentration of the corrosion inhibitor is maintained.[5]

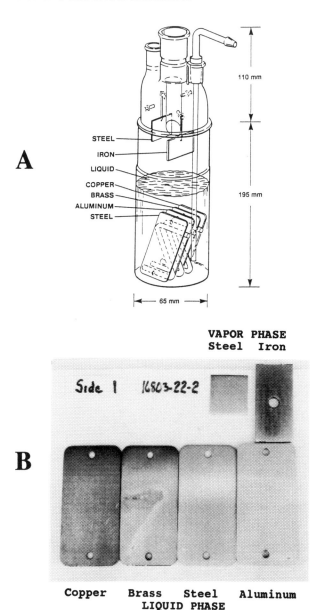

Figure 5 - Illustration of A) the corrosion test cell and B) a photograph showing the vapor and liquid phase corrosion inhibition properties achieved with an amine-containing additive.

4. Phenol Content

Increasingly stringent regulation of allowable wastes into water emitted from aluminum die casting plants has caused the aluminum die casting industry, a major user of fire-resistant hydraulic fluids, to limit the incoming "phenol content" of their hydraulic fluids. Hydraulic fluids are emitted from the plant due to fluid leakage from the hydraulic system during use. The idea is that if there are no "phenolics" contained in the incoming fluid, then they should not be present in the waste water, since no phenolics are used in the plant.

One of the specified analytical procedures for determination of phenols and phenolic derivatives, according to the Federal Register (40 CFR 403 and 40 CFR 464), is the 4-aminoantipyrene (4-AAP) colorimetric titration procedure, which is illustrated in Figure 6. The experimental procedures for this method are available in ASTM D-1738-80.

In this procedure, 4-AAP is reacted with a phenolic substrate to yield a highly colored dye which is quantitatively analyzed by visible spectroscopy at 460 nm. Since this is a colorimetric procedure, any substrate capable of reacting with 4-AAP and producing a colored product will cause an interference and produce erroneous phenol content values. Therefore, this test has been called the "phenol response" test in the industry, since the analysis is based on the formation of a colored dye and is not necessarily indicative of the actual phenolic content of the fluid. Although some fluid formulations did contain very low levels of "phenol response", in many cases there were no actual phenolic derivatives and false interference "phenol response" results were obtained. Nevertheless, it was necessary to assist the die casting industry to comply with the federal waste water requirements. This necessitated the development of fluids that both contained no phenolics and that do not produce any unacceptable "phenol response" values to the 4-AAP analysis.

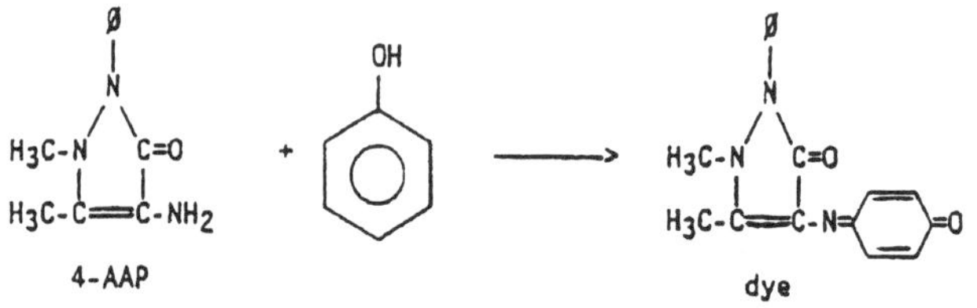

5. Low Molecular Weight Acid Formation

Although W/G hydraulic fluids have been used in the fluid power industry for 50 years, it was not generally known that low molecular weight carboxylic acids may be potentially formed during prolonged thermal abuse, such as that caused by a failed heat exchanger, could lead to increased pump wear. In a recent study, it was shown that although various carboxylic acids may be formed, only the presence of formic acid at concentrations of $>0.15\%$ would produce significant increases in wear as shown in Figure 7.[5] On the basis of this work, it is clear that, as with any class of hydraulic fluid, it is important that the fluid be used within the recommended operating limits and that the fluid be well maintained to obtain optimal antiwear results. If these procedures are followed, a W/G hydraulic fluid should not produce any significant levels of formic acid.

B. Fire-Resistance Testing

Even after fifty years of use,[2,17,18] there is still no consensus on the best evaluation procedures to quantify the relative fire resistance offered by a hydraulic fluid.[19] However, Factory Mutual Corporation has recently developed a testing procedure which provides more adequate discrimination between various

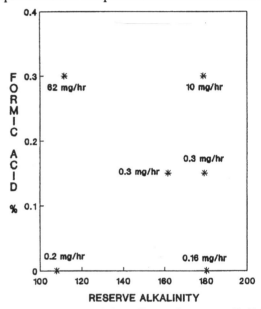

Figure 7 - Illustration of the effects of reserve alkalinity and formic acid on hydraulic pump wear rates.

fire-resistant hydraulic fluids. A summary of the results reported to date is provided in Figure 8. The order of fire-resistance offered by the various major classes of fire-resistant fluids is: water-glycol > phosphate ester > polyol ester > mineral oil. Previously, all of the fluids, except for mineral oil, received the same fire-resistance rating by Factory Mutual Corporation.

C. Performance Testing

Perhaps the most striking feature of W/G hydraulic fluids is the progressive performance improvements that have been made since their introduction in 1947. Table 2 illustrates the progressive reduction in wear rates, approaching those obtained with mineral oil, achieved with W/G fluids containing $>35\%$ water from 1955 to 1974 when tested according to

Figure 6 - Illustration of the 4-AAP quantitative dye formation reaction.

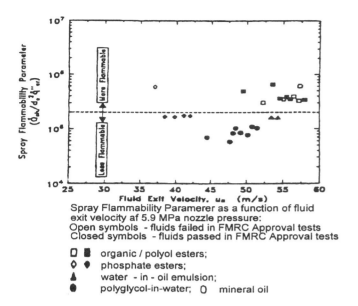

Spray Flammability Paramerer as a function of fluid exit velocity af 5.9 MPa nozzle pressure:
Open symbols - fluids failed in FMRC Approval tests
Closed symbols - fluids passed in FMRC Approval tests

□ ■ organic / polyol esters;
◇ ◆ phosphate esters;
▲ water - in - oil emulsion;
● polyglycol-in-water; ○ mineral oil

Figure 8 - Characterization of hydraulic fluid fire resistance using the Factory Mutual spray flammability parameter.

DIN 51,389 E (Vickers V-104C vane pump for 250 hours at 10 MPa (1500 PSI) and 1500 rpm.[20] Some W/G fluids today exhibit less than 100 mg total wear over the 250 hour test period.[24] It is difficult to obtain much historical data for now-obsolete formulations and the additives are no longer commercially available since the test conditions were so significantly different. For example, lower pressures and longer times and, in many cases, different pumps were used.

Table 2

Historical Wear Rates of Water-Glycol Hydraulic Fluids (1955-1974)

Year	Total Wear (mg)[1]
1955-1965	4700 - 15100
1965-1970	1300-2800
1970-1974	300-1100
after 1974	approx. 100
(Mineral Oil)	<100

1. All tests conducted according to DIN 51,389 E using a Vickers V-104C vane pump for 250 hours at 10 MPa (1500 psi) and 1500 rpm.

In another more recent study comparing the ASTM D-2882 Vickers V-104 vane pump test, 100 hours at 13.7 Mpa (2000 psi) and 1200 rpm, showed that it is now possible to formulate high performance W/G hydraulic fluids[11] that exhibit even lower wear rates than achievable with mineral oils. The reported wear rate data are provided in Table 3.[12]

Table 3

Comparison of ASTM D-2882 Vane Pump Test Results for Various Fire-Resistant Hydraulic Fluids and Mineral Oil

Fluid	Wear Rate (Mg/hr)[1]
Typical W/G Hydr. Fluid	0.65
Phosphate Ester	0.05
Polyol Ester	0.10
High Perf. W/G Hydr. Fluid	0.10
Antiwear Oil	0.24

1. These tests were conducted over 100 hours at 13.7 MPa (2000 psi) and 1200 rpm using a Sperry-Vickers V-104 vane pump according to ASTM D-2882. The pump was equipped with a 30 l/min (8 gpm) ring. The "pass" criteria is 1.0 mg/hr.

In addition to exhibiting unusually low wear rates, these high performance W/G fluids may also be used at high pressures (>34 MPa, 5000 psi), previously unattainable with other water-containing hydraulic fluids.[12]

Cole has recently shown that W/G hydraulic fluids may be used in hydraulic equipment equipped with servo valves.[21] Although this study showed that the dynamic response was slightly reduced with W/G's, the performance could be restored by using slightly larger piping, controls and servo valves.

One of the areas that has caused many problems in expanding the use of W/G hydraulic fluids in the past, was the reduced roller bearing lifetimes obtained with these fluids due to fatigue failure.[22,23] However, more recent studies have identified improved bearing materials and designs which are more suitable for use with W/G hydraulic fluids, thus permitting there use under more stringent lubrication conditions, e.g. higher pressure and greater rpm speeds.[24] This remains an active area of current research.

CONCLUSIONS

W/G hydraulic fluids were first commercially introduced in 1947 because their fire-resistant properties were critically necessary to reduce the potential for fires when hydraulic lines were ruptured on-board a naval vessel. The use of this class of fluids has grown today to include a vast array of industries, such as off-highway mobil vehicles, steel and aluminum mills, die casting and many others.

Over the years there has been many evolutionary changes in W/G fluid formulation chemistry. The nitrite containing corrosion inhibitors were replaced by non-nitrite generally amine containing inhibitors. The die casting industry was assisted in achieving federal clean water standards by the formulation of hydraulic fluids with reduced "phenol response".

Studies have been performed which demonstrate that formic acid may reduce the antiwear properties of W/G hydraulic fluids. However, it has also been shown that if hydraulic equipment is properly maintained, the resulting thermal abuse of the fluid should not occur, making this a relatively low probability occurrence.

W/G hydraulic fluids used today typically exhibit substantially lower wear rates than achievable when they were first introduced. Advances in formulation chemistry has permitted the use of W/G fluids at operating pressures heretofore unattainable (34 MPa, 5000 psi).

REFERENCES

1. E.C. Brink, Jr., *Lubrication*, **58**, (1972), p.77-96.

2. J.E. Brophy, V.G. Fitzsimmons, J.G. O'Rear, T.R. Price, and W.A. Zisman, *Ind. and Eng. Chem.*, **43**, (1951), p. 884-896.

3. F.H. Roberts and H.R. Fife, U.S. Patent 2,425,755 (1947).

4. W.H. Millett, *Appl. Hydr.*, June, (1957), p. 124-128.

5. G.E. Totten, R.J. Bishop Jr., R.L. McDaniels, D.P Braniff and D.J. Irvine, *SAE Technical Paper Series*, Paper No. 941751, Sept. 12-14, 1994.

6. A.G. Papay, *Synth. Lubricants and High-Performance Functional Fluids*, **48**, (1993), p.427-52.

7. J. Reichel, "Fluid Power Engineering with Fire-Resistant Hydraulic Fluids - Experiences with Water Containing Fluids", Presented at the STLE Annual Meeting, May, 1994, Pittsburgh, PA.

8. W.A. Zisman, J.K. Wolfe, H.R. Baker and D.R. Spessard, U.S. Patent 2,602,780 (1952).

9. Cosmetic, Toiletry and Fragrance Association, Inc., " Minutes of Nitrosamine Task Force Meeting", March 30-31, 1977.

10. W.E.F. Lewis, U.S. Patent 4,434,066 (1984).

11. W.E.F. Lewis, U.S. Patent 4,855,070 (1989).

12. G.E. Totten and G.M. Webster, *Proc. of 46th National Conf. on Fluid Power*, (1994), p.185-194.

13. H. van Oene, *SAE Trans.*, **82**, (1973), p.1580.

14. P.P. Zaskal'ko, A.V. Mel'nikova, O.N. Diment, S.G. Titurenko and E.V. Stepanova, *Chem. and Tech. of Fuels and Oils*, **10(3-4)**, (1974), p. 307-309.

15. O. Isaksson, *Wear*, **115**, (1987), p. 3-17.

16. G.E. Totten, R.J. Bishop Jr., R.L. McDaniels and D.A. Wachter, Presentation at the SAE Earth Moving Conference, Peoria, IL, April 1995.

17. R.A. Onions, "An Investigation Into the Possibility of Using Fire-Resistant Hydraulic Fluids for Royal Naval Systems",

18. G. Blanpain, "The Use of Polyglycols in French Coal Mines", *Inst. Petroleum Symp. on Performance Testing of Hydraulic Fluids*, Ed. R. Tourret and E.P. Wright, Heydon and Son Ltd.-London, October (1978).

19. G.E. Totten and G.M. Webster, *SAE Technical Paper Series*, Paper No. 932436, September (1993).

20. K.D. Aengeneyndt and P. Lehringer, *Giesserei*, **65**, (1978), p.58-63.

21. G.V. Cole, "Investigation Into the Use of Water-Glycol as the Hydraulic Fluid in a Servo System", AERE Harwell - Engineering Projects Division, July, 1984, AERE R.11324.

22. D.V. Culp and R.L. Widner, *SAE Technical Paper Series*, Paper No. 770748, 1978.

23. P. Kenny, J.D. Smith, and C.N. March, "The Fatigue Life of Ball Bearings When Used With Fire-Resistant Fluids", *Inst. Petroleum Symp. on Performance Testing of Hydraulic Fluids*, Paper No. 30, October, 1978.

24. J. Reichel, personal conversation with G.E. Totten, May 1994.

A Market Research and Analysis Report on Vegetable-Based Industrial Lubricants

Ronald A. Padavich and Lou Honary
University of Northern Iowa

Abstract

This report is the result of market research and analysis conducted at the University of Northern Iowa and commissioned by the Iowa Soybean Promotion Board. The main purpose of the study was to assess the size and scope of the vegetable oil industrial lubricant market, to identify trends within the industry, and to determine potential future regulations that would have an impact on the demand for such lubricants.

The results of the survey indicated that the need for vegetable-based lubricants is long-term. Also, there is a trend toward environmentally friendly lubricants for special applications such as mining, forestry, marine, wetland construction, government fleets and agricultural industries.

Introduction

The Iowa Soybean Promotion Board (ISPB) commissioned the University of Northern Iowa (UNI) Rural Business Expansion Program--Market Development Program (RBEP--MDP) to conduct a market research and analysis on the industrial uses of vegetable oils. Specifically, the study focused on the size and scope of the market for industrial lubricants, trends in the industry, and any potential legislation and its impact on the market place. To meet the project goals, RBEP--MDP conducted research that included: literature searchers, telephone surveys, and gathering of results from previous surveys conducted by other organizations. Conclusions and recommendations for future work were presented to ISPB on the basis of the survey results and analysis.

Size of the Industrial Lubricants Market

THE U.S. MARKET -- GROWTH AND TRENDS

The basic U.S. market for lubricants is expected to grow at an annual rate of 1.15 percent from 1996 to the year 2000, according to the latest available data from the Freedonia Group, Inc., a market research company in Cleveland, OH. They forecast the consumption of lubricants, to reach 2.72 billion gallons, valued at $7.3 billion in the year 2000, up from 2.39 billion gallons valued at $6.2 billion in 1996. Lubricant demand includes: automotive; industrial; and grease.

Below are materials obtained from the two documents: Freedonia Group Inc. and the National Oil Refiners Association. Complete texts of their reports can be obtained from their addresses appearing in the bibliography section of this report.

The U.S. market for industrial lubricants, specifically, is expected to grow at an annual rate of 1.24 percent from 1996 to the year 2000. The consumption of industrial lubricants will reach 1.16 billion gallons in the year 2000, up from 1.105 billion gallons in 1996. Industrial lubricants include: general industrial oils; process oils; industrial engine oils; and metalworking fluids. Industrial lubricant demand as a percentage of total lubricant demand will increase slightly from 42.5 percent in 1996 to 42.6 percent in the year 2000.

General industrial oil demand is expected to grow slowly from 0.411 billion gallons in 1996 to 0.414 billion gallons in the year 2000. General industrial oil demand, as a percentage of total lubricant demand, will fall over the same period from 15.8 percent to 15.2 percent. Hydraulic oil demand, specifically, is expected to increase from 0.235 billion gallons in 1996 to 0.240 billion gallons in the year 2000, an annual increase of only 0.53 percent. Hydraulic oil demand as a percentage of total lubricant demand is expected to fall from 9.0 percent in 1996 to 8.8 percent in the year 2000. General industrial oil demand includes: hydraulic oils; turbine oils; gear oils; and other general industrial oils. The overall market is expected to grow at subpar rates because lube suppliers responded very efficiently during the last recession. Producers focused on reducing costs, engineering, and other approaches to make oil last longer. They have convinced end-users to buy better oils which has lead to reductions in disposal costs. Improved recycling programs will also contribute to slow growth in the industry.

The economy has turned, but the short-term demand for lubricants and hydraulic fluids remains laggard. This is because buyers simply have not purchased as many conventional lubes as they once did. Lube buyers are more conscientious than ever and, from their suppliers, they are demanding more cradle-to-grave service, i.e. fluid management programs. Technical expertise, custom engineering, responsiveness, just-in-time delivery, and assistance with environmental and safety regulations all are in demand from some lube and fluid buyers today.

Lube distributors and manufacturers are responding. They're coming into plants and setting up and overseeing programs that will dispose of all used lubes and fluids according to local ordinances. They're monitoring preventive and predictive maintenance programs. They're adding distribution centers and placing warehouses near user locations, even overseas. Also they're developing specific applications for lubes through the advice of buyer teams.

Industrial oil and fluid consumption by the chemical and allied products industry is expected to be the fastest growing end-use market. There has been some growth in recycling and re-refining operations, which are slowing the demand for virgin base oils. Other reasons for lackluster sales include the use of electronic controls, increased use of plastics and other self-lubricating materials, recycling, and other longer-lasting lubricants. Of the industrial lubricants, synthetics should see more growth than other types of lubricants. The reasons include: performance advantages over base oil-derived lubes, including outstanding flow characteristics at low temperatures, stability, non-flammable, and improved wear and resistance.

Engine oils, which cut across all three market segments, represent the largest product area. These are followed by process oils and general industrial oils, which include hydraulic fluids, gear oils and compressor refrigeration oils; followed by other automotive fluids such as transmission fluids and gear oil, and other lubricants such as grease, which also cut across all three market segments.

Petroleum-based mineral oils have dominated the lubricant consumption market for years. Vertically integrated, multinational corporations control lubricant production accounting for 60 percent of U.S. base oil production.

Overall, worldwide lubricant supply will be more than adequate to meet demand requirements for at least the next five years. At the same time, the cost of lubricant refining is on the rise, driven by the need for higher quality stocks, while capital resources are being directed toward refinery upgrading and meeting environmental requirements by refineries.

Synthetic lubricants currently account for only one to two percent of the total lubricant market with annual growth averaging two to four percent over the next several years. Environmental concerns have sped up the arrival of synthetic oils. However, environmental concerns/regulations appear to be subsiding. Synthetics are said to last longer, resulting in less disposal requirements. Considerably more expensive, synthetics cost on average four to eight times as much as mineral oil base stocks. The higher cost of synthetic lubricants can be justified in cases where regulatory, environmental, or toxicity considerations warrant its use.

The general move toward more environmentally friendly products has been slow. A trend toward more biodegradability is part of the issue. The general order of biodegradability for common lubricants (in decreasing order) is: vegetable oils (most degradable), synthetic esters, mineral oils and alkylbenzenes, PAGs, and PAOs (least degradable). Gains in the use of vegetable oil use is generating interest in such industries as food processing, logging (i.e. chain saws), and marine (i.e. two-cycle outboard motors). Other areas include industrial hydraulic systems, which tend to leak and don't require long life or high temperature resistance.

Factors that negatively impact vegetable oil market advances include cost. Due to start up expenses and low volume; so far, environmentally aware fluids cost two or three times as much as mineral-oil fluids. Prices should decrease as more users select these products; and when biodegradability is essential, any higher cost obviously must be borne. Another factor is lack of desired oxidative stability and the subsequent polymerization of the used oil. Solutions to this problem have been reported by the author and others in earlier papers that appear in the bibliography section of this paper. Finally, the high pour point of untreated vegetable base oils limits uses in some applications.

THE WORLD MARKET -- EXPORTS, IMPORTS, AND TRENDS

The National Petroleum Refiners, National Lubricating Grease Institute, and the American Petroleum Institute do not report sales information on lubricating oils outside of the U.S. One source did suggest that the lubricants market sizes of Germany and France were each about 25 - 30 percent of the U.S. market.

As for environmental regulations, Germany is clearly the most aggressive country in the world. It appears that Switzerland and Austria are closely following the same course of action as that being pioneered by Germany. Vegetable oils and synthetics are more common there than in any other region. This is demonstrated by the fact that industrial equipment manufacturers are now offering a choice of factory-fill fluids to purchasers of heavy equipment in Europe. Sweden and other Scandinavian countries rely heavily on biodegradable synthetic oils, partially due to the low pour point problems associated with vegetable oils.

Overall consumption of vegetable oils worldwide grew at an average annual rate of 4.2 percent over the past decade. Since the 1950s, soybean oil has been the leading vegetable oil in production and in use worldwide. The U.S. soybean industry drives both the U.S. and the world oilseeds markets. But the U.S. share of world markets for soybeans and products has eroded significantly since the 1970s. Expanding oilseed production in South America, the European Union (E.U.), China, India, Malaysia, and Indonesia increased competition for U.S. oilseeds and oilseed products in the 1980s. The U.S. and the E.U. continue to be the largest consumers of vegetable oils, accounting for approximately 1/3 of world consumption. Research on the use of soybean oil as a base oil for various industrial lubricants, in the U.S., has been underway in a growers

funded project at the University of Northern Iowa - Ag-Based Industrial Lubricants Research Program.

The Oil and Gas Journal, June 6, 1994 v92 n23 p34(3) reports the exports, imports, and trends within the world market for petroleum products and renewable energy sources from the 14th annual World Petroleum Conference in Stavanger, Norway.
At this conference in Stavanger, speakers from throughout the world addressed future demand for oil and gas resources. Highlights of the picture painted by several key speakers follow:

> Peter Davies, chief economist for British Petroleum Company PLC, expects demand to continue to rise, but not at levels the industry experienced in the 1970s. He also believes oil production in the Former Soviet Union will continue to decline. A. Gharani, affiliated with OPEC's energy studies department in Vienna, expects oil demand to be 79.17 million b/d.
>
> Economic growth will steadily increase world demand for oil and natural gas during the rest of the 1990s. After the year 2000, real oil prices could jump as productive capacity expansions are needed to meet steadily growing product demand.
>
> In the very long term--2010 to 2020 and beyond--demand will depend on the extent of political and economic liberalization's and the response of governments and consumers to those changes. If that period finds a world of increased regulation and taxation, government mandated conservation, and support for nuclear and other energy sources, it could be a difficult time for international oil and gas companies.

Forces driving oil markets outside the former Soviet Union (FSU) since the mid-1980s will remain dominant through the 1990s. Oil consumption will continue to "rise," and OPEC will be the main incremental supplier. Beyond 2000, those trends will be "critically affected" by policy responses to concerns over global warming and transportation, including the role of alternative fuels.

Outlook: The Former Soviet Union

Peter Davies in The Oil and Gas Journal, (1994) indicated that oil production in Russia has fallen 3.5 million b/d since the peak in 1987. In the rest of the FSU, outlook has remained flat. High production rates have reduced ultimate recovery from many fields, and in western Siberia during the mid-1980s average well production was dropping. Without adequate investment, the trend accelerated.

Average production fell from 300 b/d/well in 1985 to an estimated 100 b/d/well in 1992, according to Davies. Water cut increased 4%/year and is now estimated at more than 75%. The energy picture in the FSU is too complex to draw simple conclusions about its future, largely because energy balances among republics and economic and political developments differ widely. But Peter Davies offered these broad conclusions:

> * Continued declines in energy and oil consumption are likely. Energy price reform "if and when it is fully implemented" will depress energy demand even further.

> * The biggest energy crisis is the decline in Russian oil production, "which is showing no signs of stabilizing."

> * Declines in coal and gas production are less than in oil.

> * Oil trade in the FSU has slumped as Russia has tried to increase hard currency exports.

Oil demand through the first decade of the next century will be influenced by economic growth, environmental legislation, and conservation. Therefore, predicting demand levels to 2010 is difficult.

HYDRAULIC OIL

A new trend towards the use of environmentally acceptable (EA) fluids appears to have emerged for hydraulic applications in the construction, mining, forestry, marine, wetlands construction, government fleets and agricultural industries.

An example is the logging industry, where there has been a switch to "green", meaning vegetable-based lubricants, because these oils degrade much quicker than petroleum. The same holds true in marine applications, for example, outboard motor lubricants. If the oil spilling into the water is vegetable-based, it degrades much quicker than the petroleum product will.

Leakage used to be an accepted part of using hydraulics and was generally ignored. Now there is growing awareness that hydraulic oil spilled into water, wetlands, and other sensitive areas can be environmentally damaging, and the EPA and other regulatory bodies are cracking down on the offenders.

Ester-based EA fluids readily degrade and lubricate better than vegetable oils. However, high cost limits use to severe applications. Vegetable oils, on the other hand,

are becoming more important because they are plentiful and cost less than synthetic fluids. Typically rapeseed-oil based, the fluids have excellent lubricating properties and work well in hydraulic systems. Ready biodegradability makes this type of fluid an attractive candidate to replace mineral oils in sensitive applications.

At three times the cost of conventional mineral-oil based fluids, EA fluids are generally considered only in environmentally sensitive areas. However, growing environmental awareness coupled with more stringent regulatory enforcement has many hydraulics users looking to these fluids. Likewise, the Army Corps of Engineers and other agencies now require EA fluids on many jobs. This added demand may bring prices down.

METALWORKING LUBRICANTS

A new generation of synthetic metalworking lubricants have been introduced in the market in compliance with environmental and technical considerations. These new lubricants are slowly gaining acceptance over the traditional oil-based formulas, which have proven ineffective in dissipating heat, prompting manufacturers to add water to such coolant formulas. Synthetic fluids have also proven effective in terms of lubricity and tool life.

Lubricity and tool life used to be the primary criteria in selecting cutting fluids. But environmental and other concerns have so increased the downstream costs associated with traditional fluids that buyers must now consider a full array of trade-offs prior to spending money.

Current EPA regulations have increased the costs users encounter when disposing of oil-based fluids. But while oil provides the lubricity required in metalworking applications, it's not so effective at dissipating heat. Water is added to coolant formulas but, here too, disposal considerations add costs for users.

Synthetic fluid performance with regard to productive tool life is approaching that of oil-based fluids in routine applications. In end milling, for example, where there is plenty of room for liberal use of the fluid, synthetics hold up quite well. Some deep hole drilling and tapping operations, however, still require oil-based formulas.

In themselves, vegetable oil-based fluids are not new. But several firms have introduced second generation products which, if they perform as advertised, might well find room in metalworking.

One supplier blends six oils to cover a wide range of industrial applications, from machining aluminum and wood to plastic and steel. This vendor goes so far as to guarantee 25 percent less consumption when compared with any pure vegetable oil, and 50 percent less consumption when compared with synthetic oil.

Distributors from around the country were skeptical when presented with these claims. None acknowledges having much experience with vegetable oil-based lubricants, but all expressed a willingness to field test formulas.

An analysis of each region's top five metalworking markets reveals that these users alone, bought more than $950 million worth of high-speed steel and carbide cutting tools and cutting fluids in 1993.

OIL FOR TWO-CYCLE ENGINES

As environmental regulations tighten there will be an increased need for environmentally safe two-cycle engine oil especially in marine (outboard motors) and logging (chain saws) applications. This view was reflected by people from all areas of the industrial lubricant's field.

INK AND DYE

Ink and dye manufacturers have seen a substantial upturn in the market in 1994, as strength in the paper industry and the general economy has made for new colorant demand. However, many say the upturn is not substantial enough to make up for the poor business conditions which plagued the market throughout the early 1990s. Of the four main ink types, growth rates vary considerably.

Petrochemical-based inks face continuing pressure due to perceived environmental impact. Water-based inks have yet to duplicate the performance of solvent-based products. For makers of vegetable oil-based inks, business has slowed considerably after a period of rapid growth beginning in the late 1980s, when only three or four manufacturers marketed a soy-based product. Energy-curable, 100 percent solid inks are said to be growing in use. These inks, which use ultraviolet light and electron beams for curing, are valued for their environmental benefits in eliminating solvents and reducing emissions.

Other Vegetable-Based Lubricants in the Market Place and Under Development

RAPESEED (CANOLA)

The rapeseed (canola oil) coalition is taking strides forward to develop vegetable-based lubricants. The Alternative Agricultural Research and Commercialization (AARC) Center, a branch of the U.S. Department of Agriculture, for example, has invested funds in a project designed to turn rapeseed oil into a major industrial feedstock. The project's goal was to develop efficient procedures for turning rapeseed oil into a low-molecular weight telomer that would have wide applications as the raw material for manufacturing lubricants and new industrial products such as high-strength nylon 1313.

Mobil Corporation has already developed and is marketing their own rapeseed oil-based hydraulic oil under the product name Mobil EAL 224H. Smaller "specialty houses" also market EA hydraulic oil, and companies such as Shell and Texaco sell products in Europe. However, these fluids are not identical. They vary widely in biodegradability, toxicity, performance, and even in the way the fluids are tested to meet these specifications. A major U.S. tractor manufacturer is marketing a biodegradable hydraulic oil, for use in ag-equipment hydraulic systems where readily biodegradable and nontoxic fluids are required.

SAFFLOWER

A member of the thistle family, safflower is also being exploited for its value in industrial niches. During the fuel supply crisis of the 1970s railroads conducted research into substitute or extender fuels. They found that a relatively small proportion of safflower oil mixed with diesel fuel reduces smoke and other particle emissions. Locomotive engines are candidates because their slower-turning engines are more tolerant of vegetable oils. The various goals for the safflower oil coalition in the state of Montana, where much of the research is being conducted, include: displacing petroleum; developing a unique new oleochemical feedstock; gaining exclusive rights to a patented, high-yield variety oilseed; improving human and animal foods; and finding new customers for a non-subsidized, drought resistant crop grown on otherwise-idle, fallow wheat land.

SUNFLOWER

Long used in the food industry, sunflower oil is gaining adherents in the cosmetic business, as well as in the industrial lubricant's field. Mud additives formulated from water-soluble combinations of silicon, phosphorus, aluminum, and boron have replaced conventional thinners and lubricating agents in Russian drilling operations. These environmentally safe additives have been found to both thin the mud and improve the mud inhibition level.

On field tests in the Tyumen area of Russia, wells drilled with these inhibitors and thinners had fewer problems with bore hole stability and sloughing formations. Common industrial wastes, such as used sunflower oil and organic acids from the food processing industry, have found an application as a base material for lubricating additives. Test wells recently drilled in the Shtockmanovskoye gas field in the Barents Sea indicate that the lubricating agents reduce well bore friction up to three times less than that in conventionally drilled wells. Use of Ultra-high oleic sunflower oils in industrial applications is increasing.

CRAMBE

North Dakota agricultural researchers are particularly interested in creating a viable market for crambe, an oilseed which, like rapeseed, is a member of the mustard family that yields an oil high in erucic acid.

Although crambe was introduced to the U.S. from Europe nearly 50 years ago by the Connecticut Agricultural Experiment Station, the oilseed has captured the attention of plains state producers only in recent years. At North Dakota State University's Carrington Research Extension Center, scientists continue to explore new industrial applications for high-erucic-acid oils. According to the center's information, opportunities for market expansion lie within three main categories of applications.

In one of these areas, the oils could increasingly be substituted for petroleum lubricants. An interest in this type of lubricant replacement in environmentally sensitive applications is seen by the researchers involved.

The high-erucic acid oil can also compete with petroleum in the plastics and nylon markets, where it performs acceptably as a feedstock. A great deal of research has been done in this area, but at this point, it seems the market is not economically competitive because recycled plastic is proving to be a much cheaper feedstock than vegetable oil.

A third area of exploration in the erucic oil market examines the oil's viability as a raw feedstock in food products, such as Procter & Gamble Co.'s fat substitute caprenin. At this point indicators suggest that the response to caprenin has not been particularly favorable.

Calgene Chemical Inc. has focused on expansion of its Erucical line of specialty esters. In addition to Erucical TD-13, a textile lubricant, and Erucical EG-20, an automatic transmission fluid additive, the company has recently introduced Erucical EH-26, an ethyl-hexyl erucate designed for high-temperature, high-lubricity applications. The Minneapolis-based National Sun Industries commissioned North Dakota farmers to plant 56,000 acres of the crop last year, a figure that more than doubles year before acreage. North Dakota farmers seem eager to grow this disease-resistant crop and they are hoping to expand production when increased demand warrants it.

CORN OIL

The National Corn Growers Association (NCGA) has sponsored projects to generate both economic and environmental benefits in developing a windshield washer solvent containing non-toxic ethanol made from corn grown in the U.S. There is a plan to begin national distribution of the fluid which will replace methanol-based solvents that can cause blindness or can prove fatal if swallowed. Methanol is made from petroleum, 50 percent of which is imported. To meet the nation's annual 120-million gallon demand for windshield washer solvent, 24 million bushels of corn would be required.

This plan is also being supported by the Alternative Agricultural Research and Commercialization (AARC) Center. As part of its 1993 round of repayable awards to promising new ventures, the AARC Center will provide funds to the NCGA to help launch production of the ethanol-based solvent and to market the new product. Once sales generate revenue, Aquinas will repay the public funds.

The advantages of vegetable oils for industrial lubricant purposes are given here as well as the advantages of soy oil in particular as compared to other vegetable oils.

Vegetable Oils are Biodegradable - 28 days vs. 6-9 months for mineral based oils. They are Non-Toxic - Derived from foods vs. chemicals; renewable - the source is a crop vs. a finite mineral deposit They promote self reliance as ample production capacity exists in the U.S. without importing materials.

Vegetable oils present higher flash Point than mineral oils, always a concern with flammable liquids; they also are generally safer to humans. They cost less than the synthetic lubricants (approximately 1/2)

SOYBEAN OIL

Soybean oil is the most available vegetable oil, over 6.0 billion gallons/year worldwide. It has a viscosity closer to mineral based oils (29 cSt at 40 degrees C). The flash point for soybean oil is high 325 (degrees C) - vs. 252

for sunflower oil. Soybean oil has a viscosity index of 246.

Although there are several disadvantages associated with vegetable oils, the reader should not assume that these oils are unsuitable for non-food applications. The primary driving forces to develop vegetable oils as an alternative to mineral-based oils is the growing environmental concerns and foreign oil dependence. The belief that these issues will eventually require alternative materials is universal. Some of the pronounced disadvantages of vegetable oils are:

Performance Limits:

   System pressure of 5,000 psi
   Increasing viscosity over time because of oxidation
   Currently not recommended for systems with servo valves
   The untreated oils are not recommended in systems that operate clutches or brakes (friction coefficient is low - too slippery)

   Temperature Limits: High - 160 degrees F
         (one company suggested 90 degrees F)

   Low - 20 degrees F (can go to 0 degrees for a short time)

   Incompatibility: mineral-based fluids (system must be flushed)

   Unstable - oxidatively & hydrolytically

BARRIERS TO MARKET PENETRATION

As the result of several conversations with a wide range of individuals, the primary barriers for market penetration of vegetable oils are:

* Performance: not equal to or superior to mineral-based products.

* Cost: currently vegetable oil lubricants are two to three times more than mineral based oils.

An engineer who was interviewed, summarized it well when he said, "Today vegetable-based lubricants are environmentally driven. Considering the operating performance disadvantages and infrequent close supervision by regulatory groups, there is very little motivation on the part of the user to spend the extra money for vegetable-based hydraulic fluids.

Environmental penalties must be very high to offset the current barriers of performance and cost."

Another barrier is product development. Research is active in many areas; plant genetics, process and formulation modifications, blending combinations of oils and additives, and testing the impact on equipment. There is still a lot of research required to make vegetable oils commercially acceptable. In a conversation with a major seed company, they revealed that approximately 15-20 percent of their research in plant breeding is focused on industrial oils. Over the next three years their focus in this area will increase significantly.

RECOMMENDATIONS FOR MARKET PENETRATION

For new products the market place always demands at least one of the following characteristics

         a) improved performance
         b) lower cost
         c) filling a unique market need

The current market offering does not satisfy characteristics a and b, but it does satisfy that of item c. Today the market is insignificant, and the potential is totally based on two future events: legislative mandates and successful advances in technology to make vegetable-based lubricants comparable with current products. With a comparable product available, the natural market acceptance and the legislative mandates can then be accelerated and it will not take ten years to create a market.

The primary expectations for vegetable-based lubricants are focused on environmental issues. However at the present time there is no momentum for a broad based regulatory mandate and associated penalties that would demand a change. There is a consensus that a need will develop for a non-mineral based industrial lubricant. This is not expected to mature for at least five years and probably ten. There are many isolated situations that have placed mineral based lubricants in a negative light, creating interest in a product here in the U.S. Because of some stringent regulations, in Germany particularly, the market is accelerating more rapidly in Europe.

Market penetration for soybean-based oil will occur if there is a substantial commitment to product development. It is believed this will have an impact on all three characteristics mentioned above for new products. It is an absolute necessity for the performance profile to be at least similar to current standards. At that point it will become an industrial lubricant rather than an experimental idea. The basic recommendation for market development into the industrial lubricant market would be a focused and coordinated industry effort with defined agendas and responsibilities. The coordination should be with:

a) plant genetics
b) process/formulation
c) blending of oils and additives
d) evaluating performance on equipment

Until a widely accepted industrial product is available, the industry would be well advised to focus on special industrial markets and niche markets outside of industrial lubricants. These marketing efforts should produce tangible results within 24 months. Although not complete, the following is a list of possible applications that could be promoted with less product development efforts.

SPECIAL INDUSTRIAL LUBRICANT APPLICATIONS

There may be a few market segments that are environmentally sensitive and would want to use such a product without legislative mandates. While the product is under development the following market segments may be early targets to begin using a soy oil-based product before congress would mandate action for industry in general.

Forest Service- loggers and government vehicles
National Park Service -loggers and government vehicles
Water Ways - Coast Guard & commercial vessels
States, Cities & Counties
Utility Companies
Cutting oils for metal removal
Release lubricants

NON-INDUSTRIAL NICHE MARKETS

* Dispersing agents for herbicides and insecticides
* Paint remover, especially in public areas [graffiti]
* Protective coatings for agricultural products [Fruits, vegetables, alfalfa, etc.]

CONCLUSION

(1) The basic U.S. market for lubricants is expected to grow at an annual rate of 1.15 percent from 1996 to the year 2000 at which time total consumption will reach 2.72 billion gallons valued at $7.3 billion. During the same time period, the demand for industrial lubricants will grow at an annual rate of 1.24 percent to 1.16 billion gallons in the year 2000. General industrial oil demand is expected to grow slowly from 0.411 billion gallons in 1996 to 0.414 billion gallons in the year 2000. Hydraulic oil demand, is expected to increase from 0.235 billion gallons in 1996 to 0.240 billion gallons in the year 2000, an annual increase of only 0.53 percent. Petroleum-based mineral oils comprise 95 percent of the total lubricant market. Synthetics currently comprise only one to two percent of the total lubricant market and should see the fastest growth among industrial lubricants at two to four percentage points per year for the next several years. Worldwide lubricant supply will be more than adequate to meet demand for at least the next five years.

(2) Movement toward synthetic and vegetable-based lubricants has been driven primarily by environmental regulations. Research indicates that current regulations might be reduced and that the possibility of new legislation being passed, that would affect disposal and handling of mineral-based lubricants, is at least five to ten years into the future.

(3) There appears to be a trend toward environmentally friendly lubricants for special applications in construction, mining, forestry, marine, wetlands construction, government fleets and agricultural industries.

(4) Although there is not an immediate need for large applications of vegetable-based lubricants, there appears to be a consensus that at some point in the future, there will be a need for such products. Several factors will affect the marketability of these products such as: new environmental regulations, price, and performance.

As a result of the research conducted, the RBEP -- MDP recommends that:

(1)   A substantial commitment to product development must be made so that a fully tested product will be ready when the market opportunity opens up in five to ten years.

(2)   Emphasis must be placed on developing a product that is at least comparable to current products on the market, at a reasonable price.

(3)   Until a widely accepted industrial product is available, the industry should concentrate on special niche markets as identified.

PERSONAL INTERVIEWS

COMPANY	LOCATION
Army Corps of Engineers	Rock Island, IL
Canola Growers of Canada	Saskatchewan, Canada
Cargill Corporation	Minneapolis, MN
Carrington Research	Extension Center North Dakota State Univ., Fargo, ND
Caterpillar	Peoria, IL
Iowa Department of Natural Resources (DNR)	Des Moines, IA
Environmental Protection Agency (EPA)	Kansas City, MO North Carolina
L & R Research Inc.	Bloomington, IL
Lubricant's World	Washington, DC
Lubrizol Corporation	Wickliffe, OH
Mobil Corporation	Alexandria, VA
National Institute of Science and Technology	Gaithersburg, MD
National Renewable Energy Lab	Golden, CO
Northland Products Co.	Waterloo, Iowa
Iowa Occupational Safety and Health Administration	Des Moines, Iowa
Pioneer Hy Bred Intl.	Waterloo, Iowa Des Moines, IA
Purdue University	Lafayette, IN
Rotary Lift Corporation	Memphis, TN
Sauer-Sundstrand	Ames, Iowa

Copies of the complete survey are available at the University of Northern Iowa (UNI) Ag-Based Industrial Lubricants (ABIL) Research Program.

BIBLIOGRAPHY

Chen, V. M., A. A., Baudouin, P. Ben Kinney, M.T. & Novick, N.J. (1991) Biodegradable and non-toxic hydraulic oils. Presented at the 42nd Annual SAE Earthmoving Industry Conference and Exposition.

Fredonia Group Inc. (1994) 3570 Uakrensville, CTR RD. #201, Cleveland, Hio 44122-5226

Hammond, E.G. & Nikolau, B. (1991) . Introduction of Exotic fatty acid in soybean oil. Progress report to the Iowa Soybean Promotion Board.

Honary, L.A.T. (1994). Potential utilization of soybean oil as industrial hydraulic oil. SAE Technical Paper # 941760. Warrandale, PA: SAE Publications.

Honary, L.A.T. (1994). "Soybean oil as an alternative hydraulic fluid". Fluid Power Journal. Milwaukee, WI: Fluid Power Society.

Iowa Soybean Promotion Board and Center for Agricultural and Rural Development (1993). The future of the Iowa soybean industry. Ames, Iowa State University.

Naegley, P.C. (1992). Environmentally acceptable lubricants. Lubrizol Corporation, Wicklife, Ohio.

National Petroleum Refiners Association. (1992) The 1993 Report on U.S. Lubricating Oil Sales. Washington, D.C.

National Petroleum News, Hunter Publishing Limited Partnership, 950 Lee St., Des Plaines, IL 60016. Telephone: (708) 296-0770.

Oil & Gas Journal, Penn Well Publishing Co., 1421 Sheridan Rd., Tulsa, OK 74101. Telephone: (918) 835-3161.

"Petroleum and Coal Products, 1987 Census of Manufactures", MC87-1-29A. Bureau of the Census, U.S. Department of Commerce, Washington, DC 20233.

Petroleum Intelligence Weekly, Petroleum & Intelligence Weekly, Inc., 575 Broadway, New York, NY 10012. Telephone: (212) 941-5500.

Slattenow, S. (1993). Student project report on biodegradability of vegetable oils in soil and water samples of Black Hawk County, Iowa. University of Northern Iowa.

The Oil and Gas Journal, June 6, "Steady rise in oil, gas demand ahead (Industry Overview)". 1994 v92 n23 p34(3).

The Oil Daily, The Oil Daily Co., 1401 New York Ave., N.W., Washington, DC 20005. Telephone: (202) 662-0700.

Weekly Petroleum Argus, Petroleum Argus, Ltd., 4801 Woodway, Houston, TX 77056. Telephone: (713) 622-3996.

(The following reports are published by the Energy Information Administration, and are available through the Government Printing Office outlets. For information concerning these outlets, call (202) 586-8800.)

Annual Energy Outlook 1993.

Annual Energy Review 1992.

International Energy Outlook 1993.

Monthly Energy Review, August 1993 .

Performance Profiles of Major Energy Producers 1992.

Petroleum Supply Annual 1992.

Profiles of Foreign Direct Investment in U.S. Energy 1991.

Short-Term Energy Outlook, 3rd quarter, 1993.

Trends and Forecasts: Petroleum Refining (SIC 2911) (in millions of dollars except as noted)

Automated Testing of Telescopic Hydraulic Cylinders

George Acker, Jr.
Dana Corp.

Mark W. Rigney
Sverdrup Technology, Inc.

Scott Hall
Santek Engineering, Inc.

ABSTRACT

The end use of high-pressure, telescopic, hydraulic cylinders for hoists requires that they undergo integrity testing, an exacting and laborious process. By integrating the mechanical structure and hydraulic power of the test stand with personal computer control and data acquisition, cylinder test data can be directly archived for permanent storage. Automated testing in the cylinder production process has proven to result in greater test repeatability, secure data archival, and reduced test time. The features and benefits of an integrated test stand are discussed, with general information on testing, mechanical design, and computer logic included for insight.

INTRODUCTION

Dana Corporation's Mobile Fluid Products Division produces telescopic hydraulic cylinders for hoists, dump bodies, and refuse-handling equipment. Three product lines are built and tested at Dana's Arab, AL plant: the HYCO line, the HPT line, and rod cylinders. The cylinders are essentially special production, designed for the needs of each end user.

Dana's customers rightly expect consistently sound, high-quality cylinders. End use demands that quality standards be met through testing of individual cylinders, an exacting and laborious process. Such testing confirms cylinder operation, leak integrity, and dimensions.

Dana's test requirements are further complicated by the need to test both single-acting and double-acting cylinders. Single-acting cylinders must be retracted during the test, while double-acting cylinders self-retract using hydraulic pressure. In addition, the extended length of Dana's telescopic cylinders is considerably greater than their collapsed length, with the longest extending almost 10 m from a collapsed length of 1.5 to 2 m. Such cylinders can also produce forces as large as 72.5 kg.

The wide variety in test requirements, and the need for faster, cost-effective testing with no loss in quality, demanded a unique approach. Dana teamed with Sverdrup Technology, Inc. of Tullahoma, TN and Santek Engineering of Guntersville, AL to produce an integrated cylinder test stand. The design combines the mechanical structure and hydraulic power of the test stand with personal computer (PC) control and data acquisition. The integrated stand is now being used at Dana in production, and has been proven to reduce test time by 30 percent with noted improvements in product quality.

The integrated test stand and its proven advantages are discussed here. Basic cylinder design and test requirements are explored, as well as the approach used to integrate the test stand with PC control. Productivity gains realized from computer control and data storage are highlighted, with information on the test process, mechanical design, and computer logic included for insight (Reference 1).

CYLINDER DESIGN

Telescopic cylinders are designed with several sleeves assembled concentric to each other. At the base of the sleeves a piston is fitted which acts as a bearing, a linear stop, and an internal seal. Pistons are usually 10 percent of the length of the rod or less. Exterior rod ends are capped with a gland nut that acts with the seal and bearing surface at the external end of each cylinder stage.

Figure 1 shows a typical single-acting telescopic cylinder. Connections to the hydraulic lines are on the cylinder's barrel end. Two connections are shown; in the field, one is plugged to allow the end user to position hydraulic lines in different locations. Gravity or payload is expected to close the single-acting cylinder.

Figure 1: Typical Single-Acting Telescopic Cylinder

Double-acting cylinders are similar to the single-acting type, except that the former are designed to close without the aid of gravity or payload. Figure 2 shows a typical double-acting cylinder. To close it, pressure is applied to the front side of the piston of each stage. Hydraulic line connections are normally on the rod end of the cylinder.

Figure 2. Typical Double-Acting Telescopic Cylinder

Rod cylinders are made in limited quantities at Dana's Arab plant. These cylinders are double-acting, identified by a single hollow rod. Typically the rod cylinder is 0.75 to 1 m collapsed and capable of extending to nearly twice its length. Such a rod cylinder is shown in Figure 3.

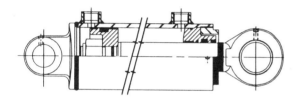

Figure 3: Typical Rod Cylinder

Two types of exterior seals are used on telescopic cylinders: "VEE" packing and "U" cup seals. Each cylinder is designed so that seal adjustments are never needed during its service life. While the "U" cup seals are self-seating, cylinder gland nuts using "VEE" packing must be torqued during testing. Tests of cylinders using "VEE" packing must be paused midway to seat the seals.

TEST REQUIREMENTS

All telescopic cylinders must be tested after assembly to confirm operation, leak integrity, and dimensions. This process is exacting and laborious, involving separate tests of each cylinder. Measurements of cylinder vibration, external leakage, cylinder "sag," proof pressure, and internal leakage must be taken, and the test results compared to database values to determine whether a cylinder passes or fails. Pass/fail measurements for each cylinder must also be stored for later reference.

It was apparent that automating this lengthy quality process would reduce test time and increase consistency in testing. Hence Dana, Sverdrup, and Santek teamed to design and produce a cylinder test stand that would be fully integrated with computer controls and acquisition.

INTEGRATED TEST STAND

The integrated test stand designed by the team consists of three separate elements: (1) the test stand structure, which includes a slave cylinder and clamping mechanisms, (2) the hydraulic supply system, and (3) the computer controls and data collection.

TEST STAND STRUCTURE - A phantom view of the test stand is presented in Figure 4. As shown, the test cylinder is fastened to a fixed-base clamp on the test stand and to a moveable rod clamp. The latter is attached to a test car that can move (on tracks) the length of the test stand. A slave cylinder is fixed to the test stand near the base clamp of the test cylinder; it passes through the test car with the test cylinder rod clamp attached, and is secured to the slave car. Since the slave cylinder produces the most force in the retract mode (collapsing the test cylinder), the slave car pushes the body of the test car during the retract cycle. The slave car is attached to the test car with a clamp mechanism that pulls the test car into the correct position for each cylinder length at the start of testing.

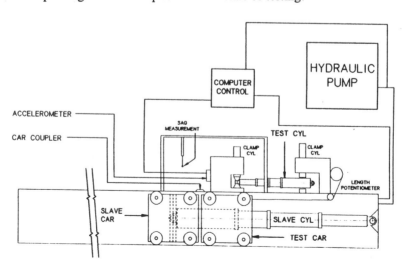

Figure 4: Phantom View of Integrated Test Stand

For all tests, the slave car positions the test car for easy loading of cylinders to be tested. Minimal repositioning of the test car is required. The test stand was also designed to accommodate future automatic loading and unloading of test cylinders.

HYDRAULIC SUPPLY SYSTEM - Figure 5 is a hydraulic schematic of the test stand. A two-stage hydraulic pump supplies oil for the test. The first-stage 2.4-liter/sec pump is set to a pressure of 10,000 kPa, allowing a smaller 0.5-liter/sec pump to pressurize the cylinders for proof pressure. The slave cylinder operates from the low-pressure pump at a constant 10,000 kPa. A 3-μ kidney loop filters the test oil continuously during operation. Cylinder length is determined by a linear string potentiometer attached to the test car and compensated for the zero position of the test car. All valves used to direct oil flow during testing are computer controlled. Return oil passes through 3-μ Pall filters.

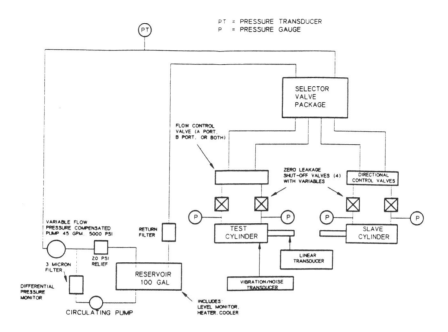

Figure 5: Hydraulic Schematic of Integrated Test Stand

COMPUTER CONTROLS AND DATA COLLECTION - The selection and configuration of both hardware and software for the integrated test stand were carefully considered, with the following design solutions developed (References 2 and 3).

The system application drove the decision to use a PC-based data acquisition and control system (DACS). The controlling PC is a Hewlett-Packard VL2 with a 486-66 microprocessor. The computer is configured with 16 MB of random-access memory (RAM), 340 MB of fixed-disk storage, a 250-MB internal tape backup, one parallel port, two serial ports, a 38-cm flat-screen color monitor, and a high-quality color printer. Because the DACS must operate in an industrial environment, a National Enclosure Manufacturers Association (NEMA) 4-rated portable, temperature-conditioned cabinet was selected to house the equipment.

The software package selected to implement the DACS was National Instruments' LabVIEW for Windows®. LabVIEW was chosen because it provides a user-friendly graphical interface, is easily maintained, and is expandable. Another off-the-shelf Windows product selected was Microsoft Excel®. Excel can accommodate all required test data handling, and the dynamic data exchange between LabVIEW and Excel is straightforward and seamless. Software operation was designed to require minimal user interaction. System navigation is accomplished using function keys, and the only data manually entered by the operator is the test cylinder serial number (for accurate data archival).

To collect data from the signals and control system valves, multiple input/output (I/O) data channels were required. These channel requirements and the smooth integration of National Instruments' data acquisition products with LabVIEW narrowed card selection to a single National Instruments' AT-MIO-16D. The card contains a data acquisition board with eight differential analog inputs, two analog outputs, and 32 digital inputs or outputs. Two stepping motor devices for cylinder sag measurements are controlled via the RS-232 serial interfaces.

AUTOMATED TESTING WITH THE INTEGRATED STAND

With the integrated stand, test execution is divided into two sequences to accommodate both single-acting and double-acting cylinders. Each test sequence is controlled by a test matrix which provides direct control for each step. Since such actions as test cylinder and slave cylinder extension and retraction, car coupling, and test cylinder clamping are controlled solely with solenoid valves, each element in the matrix represents a binary state for a solenoid valve (energized or deenergized).

Solenoid Valve Control

X	= Energized
(empty)	= De-energized

Test Step	Valve					
	Base Latch Down	Base Latch Up	Rod Latch Down	Rod Latch Up	Master Cylinder Extend	Master Cylinder Retract
1. Clamp cylinder	X		X			
2. Extend cylinder	X		X		X	
3. Retract cylinder	X		X			X
.						
.						
.						
.						
n. Remove cylinder		X		X		

Figure 6: Typical Test Matrix (Simplified)

A typical (simplified) test matrix is shown in Figure 6. Each column in the matrix is associated with a particular solenoid valve and each row represents an individual test step. Therefore, operations to be performed on all solenoid valves at every test step can be assigned. The duration of each step is largely determined by test cylinder response times. For example, one test matrix step is to extend the test cylinder. When this step (row) is executed, the digital output channels for the solenoid valves are configured as represented (columns) in the matrix. The cylinder then begins to extend. The test sequence module waits until the measured length of the test cylinder matches the extended length retrieved from the test cylinder configuration for a specified interval, and then executes the next step in the test sequence matrix. This process continues until the test is complete and the cylinder is ready for removal from the test rack.

By carefully configuring each element of the test matrix, every step of a cylinder test sequence can be completely defined, resulting in greater test repeatability, faster cycle times and, ultimately, more accurate knowledge of cylinder performance. The system also allows easy modification of test steps, to accommodate any changes in test sequence or solenoid valve requirements.

The application software's primary components execute independently yet share key sets of data. Figure 7 depicts the software application approach. Upon system startup, with no operator control, the application initiates data acquisition, alarm/limit checks, oil reservoir temperature, auxiliary oil tank level control, and the pretest mode of operation. The operator then enters the serial number of the cylinder part number, and the software retrieves all predefined test information and prepares the test stand to receive the test

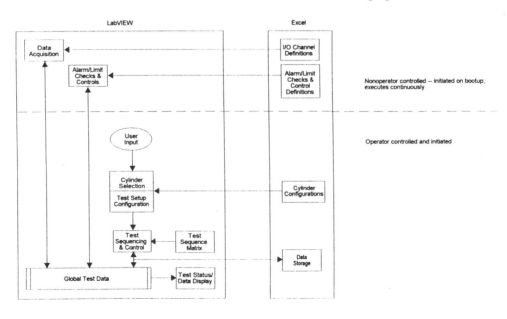

Figure 7: Application Software Approach

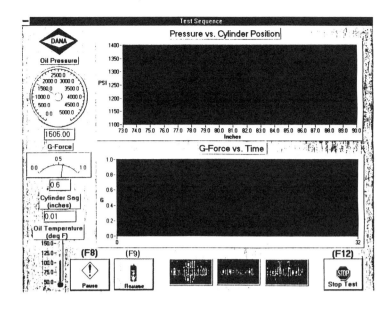

Figure 8: Typical Data Displays

cylinder. Both coupling devices are automatically opened and the test car and slave cylinder are moved to the position necessary to allow test cylinder installation. Once the cylinder is installed and both ends are clamped (through automatic control), the test sequence is initiated by the operator.

Pressures commonly seen in the field are applied to the test cylinder in three strokes. The first two are low-pressure (10,000-kPa) strokes, applied to bleed the air from the cylinder. The last stroke applies proof pressure on the fully extended cylinder, holding the pressure on the extend side of the piston for 30 sec. During this time a robotic manipulator measures the height of the cylinder at mid-span. The cylinder is then depressurized and another measurement taken. The delta between the two mid-span measurements is a measure of the sum of the tolerances of the different stages of the cylinder ("sag"). The cylinder is then retracted to the "closed" position.

With double-acting cylinders, proof pressure is applied to the retract side of the piston for 30 sec, thus testing all seals in the cylinders for leakage. The resultant pressure at the end of the 30-sec interval is compared to a pressure equal to 66 percent of the proof pressure (maximum test pressure), with the degradation representing a good measure of external leakage. Then an internal leak test is conducted by blocking the forward port and applying proof pressure to the retract side of the piston. The differential area of the rod piston allows extension of the rod if internal leakage exists. Some internal leakage is acceptable due to the clearance of the piston and cylinder.

Each cylinder is also examined for vibration during all phases of testing. An accelerometer is mounted to the rod end clamping mechanism, to test the horizontal acceleration of the rod as the cylinder is moved in the extend and retract directions.

Two data plots are displayed during the tests to allow on-line, real-time test decisions (see Figure 8). As shown in the figure, the upper plot features pressure versus stroke length, while the lower plot depicts acceleration measured in gravitational force versus time. No interaction is required throughout the test sequence, but the operator can pause the test sequence at any necessary point by pressing a function key. Resumption of testing is also handled this simply.

At the end of testing, all pass/fail options for the cylinder are identified on-screen and in a report (specifications for each cylinder tested are stored in a database, with easy access by part number). Test result values for open and closed length, maximum vibration, maximum pressure, sag, and internal and external leakage can then be compared to determine whether the cylinder passes or fails. This pass/fail data will be saved on disk and on tape by serial number for the life of the cylinder, helping protect these important records.

CONCLUSIONS

The team-designed integrated test stand became operational for Dana Corporation in late October 1994. Since that time, the stand has been operated continuously in production with no major problems. The stand has proven to be reliable and easy to maintain, while consistently improving the productivity of Dana's plant. Automation at Dana is producing significant test efficiency gains, with a proven reduction in test time of approximately 30 percent.

The advantages of automating the telescopic cylinder test process are many. Computer control allows exact test sequence definition, while retaining the flexibility to accommodate necessary changes. Automation also ensures greater test repeatability. Computerized data storage means not only more accurate recordkeeping, but also longer-term support to cylinder customers. The biggest benefit gained from automated testing, however, is a superior product. This approach represents a significant advance in quality control for cylinder manufacturers, with end users reaping the rewards.

Only after carefully defining the overall requirements can such an automated system be designed and built effectively and affordably. Maximizing the use of off-the-shelf hardware and software minimizes development and cost risks typically associated with custom-developed hardware and software solutions. In this case, innovative software design techniques simplified software development and also contributed to easy maintenance and improved performance of the integrated test stand.

REFERENCES

1.   Schweber, W. L., "Considerations in Selecting a Small Computer for the Hydraulics Lab," 851591, Society of Automotive Engineers, Warrendale, PA, 1985.

2.   Sherwin, K. A., "A Second-Generation Personal Computer Network for Automated Hydro-Mechanical Testing," 841136, Society of Automotive Engineers, Warrendale, PA, 1984.

3.   Dyvig, D. E. "Using Personal Computers in the Hydraulics Test Lab," 841137, SAE, Warrendale, PA, 1984.

A Statistical Method for Damage Detection in Hydraulic Components

F. Honarvar and H. R. Martin
University of Waterloo

ABSTRACT

The detection and tracking of the damage process between surfaces in contact, together with an estimation of the remaining service life, are significant contributions to the efficient operation of hydraulic components.

The commonly used approach of analyzing vibration signals in terms of spectral distributions, while being very effective, has some shortcomings. For example, the results are sensitive to both load and speed variations.

The approach presented in this paper is based on the fact that the asperity distribution of surfaces in good condition have a near normal probability distribution. Deviation from this can be tracked using statistical moments. The Beta probability distribution provides a number of shapes, including normal, under the control of two positive numbers, α and β. Unlike the normal distribution, which indicates defects by kurtosis values higher than 3.0, the Beta distribution provides more flexibility.

This paper discusses the results of analysis and testing carried out on damaged bearings, typical of those common to hydraulic equipment, using this approach.

INTRODUCTION

There are many examples of sliding and rolling surfaces in hydraulic components, the most common being rolling element bearings, and the interaction of gear tooth surfaces. There is a potential problem common to all types of surface contact, and that is if a failure process starts to occur, the result can be costly, both financially and in lost production.

As damage between surfaces progresses, vibrational energy is generated which is normally of a higher level than that usually measured from surfaces in good condition. This energy can easily be detected using standard accelerometers. In either condition, the energy spectrum can be broadly divided into;

(a) Signals in the range 0 to 25 kHz, which can be examined in both the time and spectral domains using readily available equipment.

(b) Signals in the range 25 kHz to 700 kHz, which are really acoustic emission type signals that need more specialized equipment.

Most conventional approaches use the first range of frequencies. Within this range are low frequency signals, result from ball passing frequencies, interaction with cages and inner and outer races. These are generally deterministic in nature. From around 4 kHz up to 25 kHz, the signals are more random in nature, and depend on the general condition of the surfaces in contact.

As an alternative approach, the British Steel Corp. in conjunction with the machinery health monitoring group at the University of Southhampton (1979)[1], developed a system based on statistical parameters which utilized the contents of the signal source to great advantage.

The acceleration amplitude distribution of the signal source, as detected using an accelerometer, can be expressed in terms of the probability density function (pdf). This is derived from that portion of time that a signal falls within a particular amplitude band. Impacts resulting from surface damage will generate larger values of acceleration amplitude at the extreme of the range; in other words, more spikes will be present in the vibrational signal. It is well established that a smooth surface will generate a pdf curve that closely approximates to a normal distribution, and any surface damage will distort this shape.

The shape of the pdf clearly holds information about the condition of moving surfaces in contact, based on the collection of time domain data. While the normal distribution has a single shape, an alternative pdf, the Beta distribution [2,3], has the flexibility of assuming a family of shapes. It also tends to minimize variations in the results due to spurious data by taking into account both peaks and valleys of the data set.

THE BETA DISTRIBUTION

If y is a random variable, representing data collected using an accelerometer, having a range from a to b in amplitude, and if α and β are numerical parameters of the beta dustribution, determined from the mean and variance of y, then the Beta probability distribution function is given by [3]

$$p(y) = \frac{1}{B(\alpha, \beta)} \; [y(b-a) + a]^{\alpha-1} \; (1 - y(b-a) - a)^{\beta-1} \tag{1}$$

where $a<y<b$, and represent maximum and minimum amplitudes in the time domain data. If the signal is normalized to unity, sensitivity to extreme values in the data is reduced, and EQ (1) reduces to;

$$p(y) \begin{cases} = \dfrac{1}{B(\alpha, \beta)} \; y^{\alpha-1}(1-y)^{\beta-1} & \text{if } 0 \le y \le 1 \\[2mm] = 0 & \text{if } y < 0 \ or \ y > 1 \end{cases} \tag{2}$$

where α, β are positive numbers that control the shape of the distribution and $B(\alpha, \beta)$ is the Beta function represented by the integral

$$B(\alpha, \beta) = \int_0^1 y^{\alpha-1} (1-y)^{\beta-1} \; dy \tag{3}$$

This integral can also be expressed in terms of the gamma function, such that

$$B(\alpha, \beta) = \frac{\Gamma(\alpha) \; \Gamma(\beta)}{\Gamma(\alpha + \beta)} \tag{4}$$

where, for example

$$\Gamma(x) = \int_0^\infty y^{(x-1)} \; e^{-y} \; dy \tag{5}$$

It has been shown [4], that any statistical moment M_k, for this distribution can be calculated from,

$$M_k = \frac{(\alpha+k-1)(\alpha+k-2)...(\alpha+1)\alpha}{(\alpha+\beta+k-1)(\alpha+\beta+k-2)...(\alpha+\beta+1)(\alpha+\beta)} \tag{6}$$

where k is the moment indices. This results in the first and second moments as,

$$M_1 = E \; [y] = \frac{\alpha}{\alpha+\beta} = \overline{y} \tag{7}$$

$$M_2 = E \; [y^2] = \frac{\alpha(\alpha+1)}{(\alpha+\beta)(\alpha+\beta+1)} \tag{8}$$

The variance is therefore expressed as,

$$\sigma^2 = M_2 - \overline{y}^2 = \frac{\alpha\beta}{(\alpha+\beta)^2(\alpha+\beta+1)} \tag{9}$$

The fourth moment is,

$$M_4 = \frac{3\alpha\beta(2\alpha^2-2\alpha\beta+\alpha^2\beta+\alpha\beta^2+2\beta^2)}{(\alpha+\beta)^4(\alpha+\beta+1)(\alpha+\beta+2)(\alpha+\beta+3)}$$

normalizing M_4 with respect to M_2 results in the beta kurtosis

$$\begin{aligned} K_\beta &= \frac{M_4}{M_2^2} \\[2mm] &= \frac{3(\alpha+\beta+1)(2\alpha^2-2\alpha\beta+\alpha^2\beta+\alpha\beta^2+2\beta^2}{\alpha\beta(\alpha+\beta+2)(\alpha+\beta+3)} \end{aligned} \tag{10}$$

The mean $\overline{y}$ and variance σ^2 can then be used to find values of α and β, where,

$$\alpha = \frac{\overline{y}}{\sigma^2} \; [(1-\overline{y}) \; \overline{y} - \sigma^2] \tag{11}$$

$$\beta = \frac{\alpha(1-\overline{y})}{\overline{y}} \tag{12}$$

Details for the computational algorithm of these relationships can be found in reference [3]. Some examples of the effects of varying α and β are shown in Figure 1.

SIMULATION STUDY

In order to obtain a feel for the type of results that might be achieved, signals that are typical of those generated by a bearing were simulated using MATLAB. For example, a Gaussian random signal was assumed to represent time domain data from a bearing in good condition; whereas Gaussian random with superimposed spikes representing a typical ball passing frequency with a defect on the outer race, represented a damaged bearing (see Figure 2).

Figures 3a and 3b show the results of such a simulation. The x-axis in each plot represents increasing damage, so that zero would represent a bearing in good condition with no spikes in the data, while a value of 20 would represent significant spiking. Note that the K_β values increase with damage intensity while the S_β values settle out to around zero.

Figure 4 represents similar results for the beta skew on simulated data that has been rectified. The decreasing trend of S_r^β is due to the fact that the data is normalized with respect to its maximum value.

EXPERIMENTAL TESTS

The experimental testing and data acquisition arrangement are shown in Figure 5. It consists of two housings, each holding sample ball bearings that are similar to those used in hydraulic equipment. A variable speed dc motor was used to control the shaft speed. Data was collected using a standard accelerometer with a resonance frequency above 50 kHz and a sensitivity of 100 mv/g. In order to get rid of the effects of low frequency vibrations, such as ball passing frequencies, and a structural resonance at 3 kHz, a digital high pass filter with a cut off frequency of 4 kHz was used. The frequency range of 4-20 kHz was used for analysis of the data, with a sampling rate of 40 kHz.

A total of six bearings of similar type were used, with a variety of faults introduced artificially. The control was one of the bearings kept in good condition. Table 1 summarizes the test conditions.

Table 1. Test Conditions Summary

File Ref	Condition	Comment
GBR	good	---
ORD	scratch, outer race	3.3 mm long by 0.5 mm wide
IRD	scratch, inner race	3.0 mm long by 0.25 mm wide
PRL	improper lubrication	lubricant cleaned out
ABR	abrasive material present	1 gr/40 mℓ, cement dust

Figure 6 shows the experimental results for a good bearing over a speed range up to 3000 rpm. The plot shows the results for zero radial loading, and a 827 N radial load. On average over a wide range of tests the beta kurtosis value for a good bearing was found to be 2.6. Figure 7 shows results for similar conditions, with a scratch on the inner race (see Table 1). Most industrial machines are driven around 1740-1800 rpm, where it can be seen that the beta kurtosis values are significantly different, for healthy and damaged bearings. Referring to the simulation results shown in Figure 3a, it is seen that these results show K_β=2.6 for a good bearing and 2.87 for a damaged one. Figures 8 and 9 show the result of increasing radial load, for two fixed speeds. Similar characteristics are observed.

Figures 10a and 10b show the experimental results obtained for damage imposed on the outer race. In this case (see Table 1) the damage was less severe, and the results are not quite so distinct as the previous case.

The simulation results indicated that little information could be obtained from tracking the skew of the beta function (see Figure 3b), however, if the raw data was rectified first, the rectified beta skew (see Figure 4) indicated significant change with damage. Figures 11a, b, c and d show the corresponding experimental results for a good bearing and a damaged bearing.

An alternative method of presenting the results of such tests is in the form of a damage map. The more conventional method of plotting speed against acceleration level results in a curve, however it can be demonstrated [5] that for a good bearing the kurtosis remain fairly constant with speed. In the case of the beta kurtosis, the results for constant speed, but varying radial load are shown in Figure 12. Compared against a good bearing normally lubricated, was a good bearing with all the lubrication washed out, so that it was

dry (PRL); and the same bearing with 1 gm per 40 mℓ of cement dust in the lubricant (ABR). This clearly detects the difference in operating conditions.

CONCLUSIONS

A wide range of different methods of tracking bearing damage are commercially available, each of which have their individual strengths and weaknesses. The statistical approach offers a different way to analyse the time domain data, which shows a great deal of promise as a means of tracking the development of damage between surfaces in contact. The approach offers a low cost tool for maintenance purposes, but also has potential for quality control applications. The simulated bearing conditions, based on the assumption that a good bearing displays a normal probability distribution, showed clear trends in both the beta kurtosis and beta skew values, as damage level was increased. This result showed good correlation with the experimental results, especially around a shaft speed of 1800 rpm, a typical running speed for many hydraulic pumps.

When the beta kurtosis moment was plotted against acceleration levels, in the form of a damage map, the results showed a significant indicator of damage. The results shown in this presentation are only a small sample of all the tests carried out, however the research is still at an early stage. There are many other probability distributions to investigate; as well as ways of processing the data.

ACKNOWLEDGEMENT

The authors would like to acknowledge the financial support of the Natural Sciences and Engineering Council of Canada, through grant No. 0GP0007729.

REFERENCES

1. Dyer, D. and Stewart, R.M. Detection of Rolling Element Bearing Damage by Statistical Vibration Analysis, *ASME, J. Mech. Des.*, Vol. 100, 1978, pp. 229-235.

2. Whitehouse, D.J. *Beta Functions for Surface Topology Annuals of CIRP*, Vol. 27/1, 1978, pp. 491-496.

3. Martin, H.R., Ismail, F. and Sakuta, A. Algorithms for Statistical Moment Evaluation for Machine Health Monitoring. *J. Mech. Syst. and Sig. Proc.*, 6(4), 1992, pp. 317-327.

4. Larson, H.J. *Introduction to Probability Theory and Statistical Interference*, J. Wiley, 1982.

5. Volker, E. and Martin, H.R. Application of Kurtosis to Damage Mapping, *Proc. 4th IMAC*, Los Angeles, 1986, pp. 629-633.

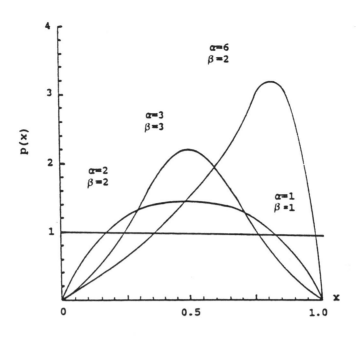

Figure 1. Effect of α and β on the Beta function

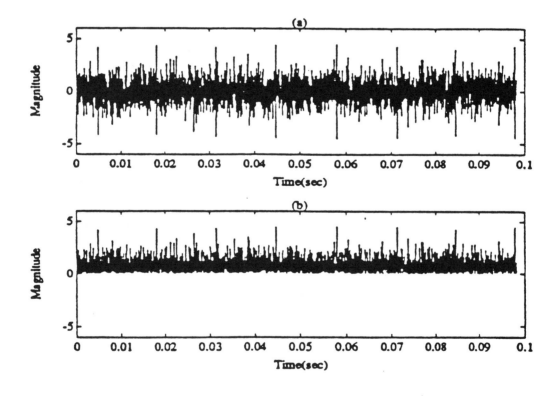

Figure 2. Data sample of a simulated damaged bearing

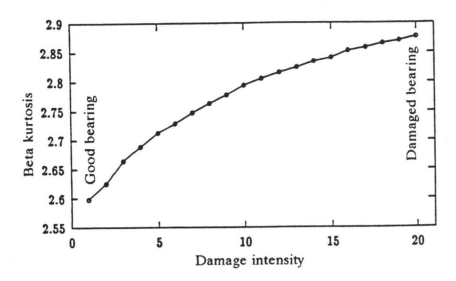

Figure 3a. Beta kurtosis values - simulated

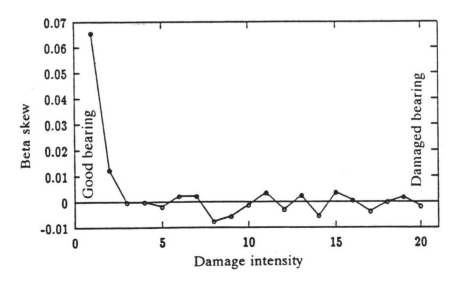

Figure 3b. Beta skew values - simulated

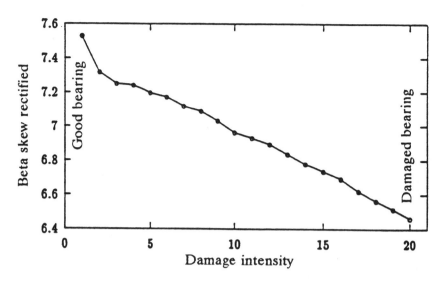

Figure 4. Beta skew values, rectified data - simulated

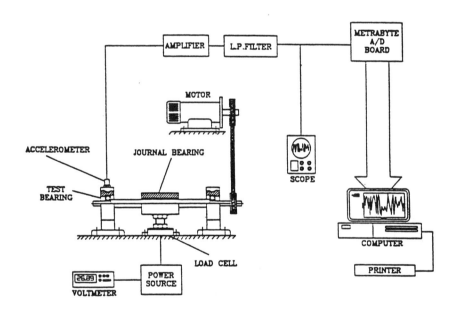

Figure 5. Experimental Rig

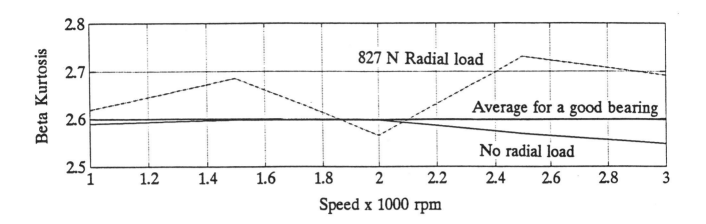

Figure 6. Beta kurtosis v speed for a good bearing(GBR)

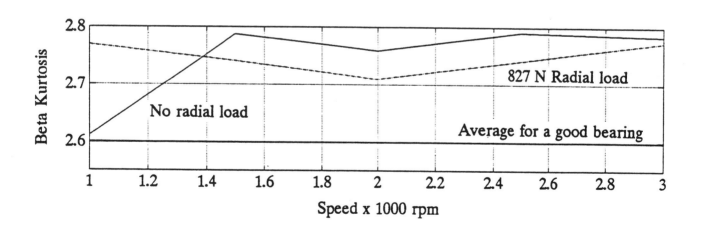

Figure 7. Beta kurtosis v speed for a damaged bearing(IRD)

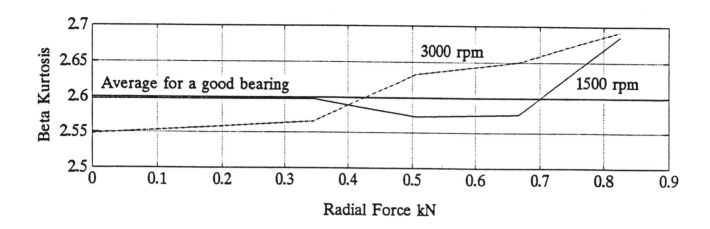

Figure 8. Beta kurtosis v load for a good bearing(GBR)

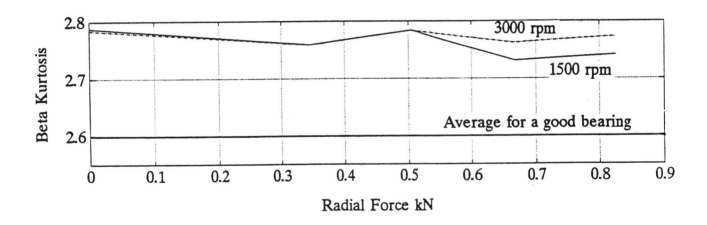

Figure 9. Beta kurtosis v load for a damaged bearing(IRD)

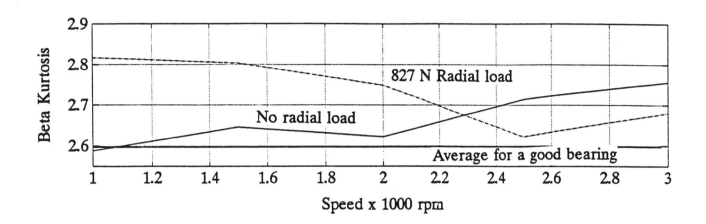

Figure 10a. **Beta kurtosis v speed for a damaged bearing(ORD)**

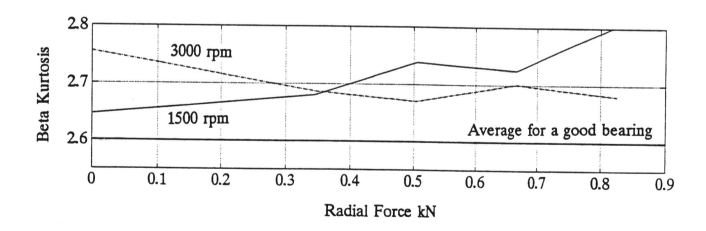

Figure 10b. **Beat kurtosis v load for a damaged bearing(ORD)**

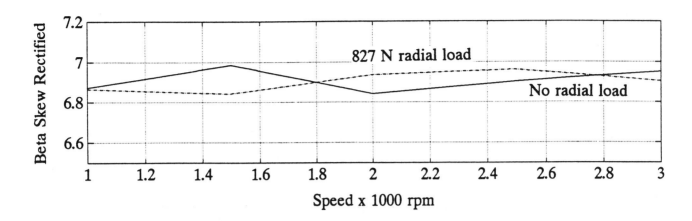

Figure 11a. Beta skew rectified v speed for a good bearing(GBR)

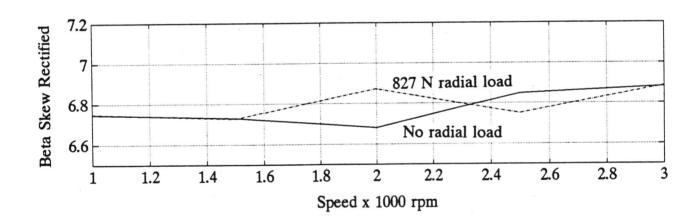

Figure 11b. Beta skew rectified v speed for a damaged bearing(ORD)

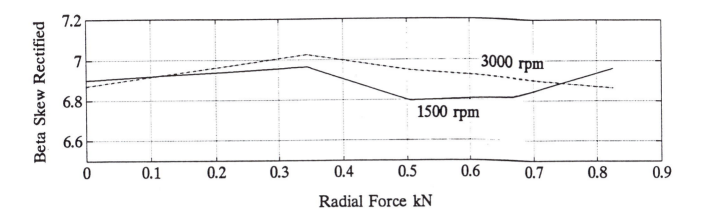

Figure 11c. Beta skew rectified v load for a good bearing(GBR)

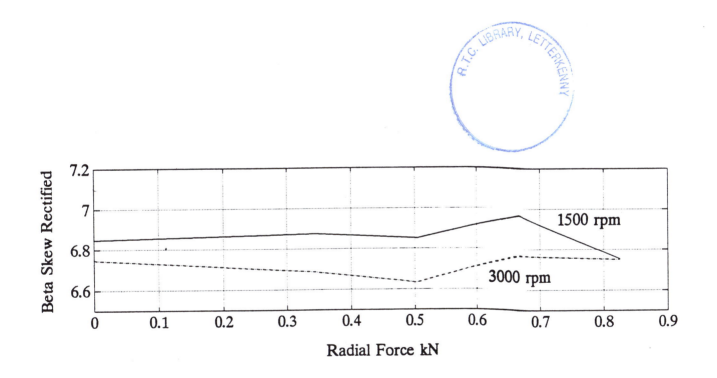

Figure 11d. Beta skew rectified v load for a damaged bearing(ORD)

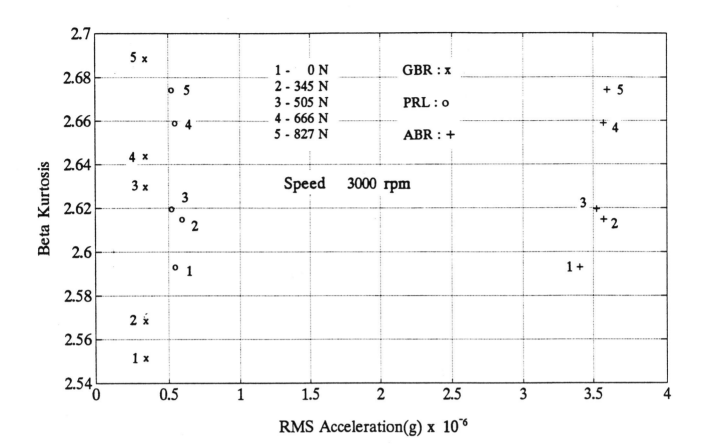

Figure 12. Damage map

952091

Performance Map and Film Thickness Characterization of Hydraulic Fluids

L. D. Wedeven
Wedeven Associates Inc.

G. E. Totten and R. J. Bishop, Jr.
Union Carbide Corp.

ABSTRACT

A new approach is presented for the evaluation of hydraulic fluids for pump wear performance. The approach uses performance maps developed in terms of rolling and sliding velocities to establish lubrication and failure regimes for test fluids. Testing pathways within the performance map can determine the fluid attributes for wear, scuffing and traction (friction). The measurement of oil film thickness with optical interferometry is used as part of a comprehensive approach for fluid evaluation. These measurements allow the lubricated contact itself to provide the viscous film forming properties of the fluid. An "effective" pressure-viscosity coefficient is determined for a range of fluid types. Performance mapping, together with film thickness measurements, provide an insight into the fundamental chemical and physical attributes of the fluid. The new approach provides an alternative to the limited reliability of bench tests and the time consuming and expensive hydraulic pump tests.

INTRODUCTION

There has been an ongoing interest in the development of "bench tests" as alternatives to hydraulic pump testing to model hydraulic fluid lubrication performance. Examples of bench tests include: Shell 4-ball, Falex pin-on-V-block and the FZG gear test. The standard ASTM versions of these tests do not adequately model lubrication performance; although, with considerable experimentation, test conditions can be identified that appear to correlate with hydraulic fluid lubrication performance in a hydraulic pump[1].

In some cases, where standard bench tests are inadequate, custom bench tests are constructed. For example, Jacobs, et. al.[5] developed a bench test to model sliding wear regimes that occur in hydraulic pumps as a function of material pairing and also fluid contamination, as illustrated in Figure 1.[5]

Instead of using one or more of these bench tests to characterize the lubrication performance of hydraulic fluids, it is also possible to evaluate a fluid over a broad range of

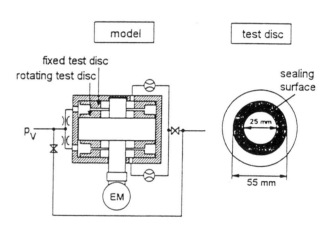

Figure 1 - Schematic diagram of the bench test used by Jacobs, et.al. to model sliding wear in hydraulic pumps.

fundamental performance variables such as contact velocity, load and fluid viscosity. Tessmann and Silva[2,3] characterized hydraulic fluid performance using the classic Stribeck-Hersey curve,[4] as shown in Figure 2.

An alternative strategy to the use of bench tests, either those commercially available or custom tests,[6] is to characterize hydraulic fluid performance with respect to fundamental lubrication performance variables which can be related to lubrication and failure modes in pump hardware. This can be accomplished by examination of the variation of lubrication performance relative to the entraining velocity (R) which generates an oil film and a sliding velocity (S) which is a major component for oil film breakdown.[6-8]

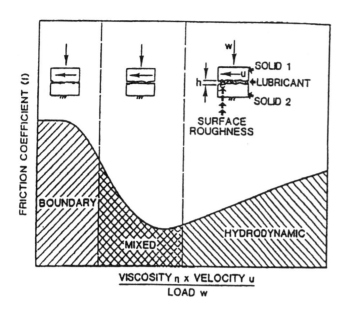

Figure 2 - Stribeck-Hersey curve illustration used by Tessmann and Silva[2,3].

Figure 3 - The WAM3 machine uses a ball-on-disk contact configuration. The normal stress, tangential strain and contact kinematics are computer controlled.

Fluid evaluation over a sufficient range of R and S velocities permits the characterization of the performance of the hydraulic fluid with respect to safe and unsafe lubrication and failure regions.

This paper describes the construction of performance maps which characterize the lubrication performance of hydraulic fluids. Experimentally measured EHD film thickness, along with an analysis to determine the viscous film-forming properties of a fluid is described. The analysis is used to derive an "effective" pressure-viscosity coefficient over a temperature range.

DISCUSSION

A. Test Methodology

1. WAM3 Test Machine - Traction coefficients and scuffing behavior of two model hydraulic fluids and two reference oils were determined using the "WAM3" test machine, shown in Figure 3[7,8]. The test machine is designed for precision control and monitoring of a single contact in three-dimensional space[6,7].

The test machine provides a tribo-system consisting of a ball-on-disk specimen configuration. The contacting specimens can be operated over a wide range of loads, rolling and sliding contact velocities, temperatures and environmental conditions. The flexibility of the machine allows simulation of specific contact conditions and performance mapping over a wide range of contact conditions.

The ball drive is positioned at an angle to control the amount of spin within the contact. The ability to control the

tribological input variables provides an opportunity for wear measurement and the development of tribological performance maps in terms of R and S. The range of R and S available with the WAM3 are shown in Figure 4. For comparison, the operating conditions for a 4-ball test, FZG test and Ryder Gear test are also shown.

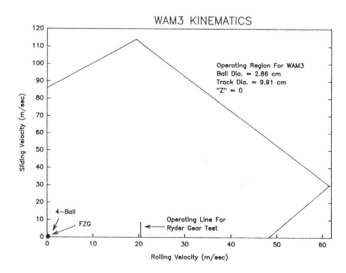

Figure 4 - Map of entraining velocity (R) and sliding velocity (S) available for the WAM3 machine.

Precise control of ball and disk surface speeds and their direction (velocity vectors) is provided. This allows the introduction of a large range of rolling and sliding velocities which can be independently controlled. The independent control of R and S provides a "decoupling" of entraining

42

velocity in the inlet region from the sliding velocity in the Hertzian region. This decoupling provides a method for separation, or at least controlling, the interaction between the physical and chemical properties of a hydraulic fluid. It also provides a direct connection to specific rolling and sliding velocities encountered across the face of a contacting gear mesh or across the contact ellipse of rolling element bearings. Heating or cooling of individual specimens provides thermal control of test conditions. Continuous recording of traction, load and sliding speed allows real time calculation of flash temperature.

2. EHD Film Thickness Measurements by Interferometry

Ueno and Tanaka calculated the film thickness of different classes of hydraulic fluids under various load conditions encountered during the operation of a hydraulic pump.[16] Their results showed that film thickness was an important parameter controlling wear in hydraulic systems.

The WAM3 machine provides the capability for the measurement of elastohydrodynamic (EHD) film thickness between the two contacting surfaces. This is accomplished with a dual chromatic optical interference fringe system. The interference fringes are calibrated to supply the optical film thickness between a transparent (pyrex) disc and a smooth M50 steel ball. A typical interference fringe pattern around the contact is shown in Figure 5.

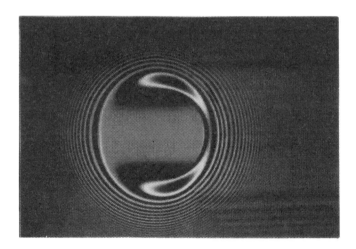

Figure 5 - Dual chromatic interference fringe pattern used to measure EHD oil film thickness between ball and disc contact.

The EHD contact is monitored with a microscope, closed circuit TV and a video recording system. The tests are conducted with a load of 44.48 N (10 lbs) which gives a Hertzian contact stress of 0.59 GPa (85,000 psi). The EHD film thickness measurements are performed by recording the optical fringe color in the center of the contact as a function of rolling velocity. Optical film thicknesses are measured at

the "center" of each fringe and at the transition between each fringe. The optical film thickness data is converted to actual film thickness by making corrections for the refractive index of the test fluid, including the effect of pressure on density under the Hertzian contact. The Lorenz equation[17] is used to correct the refractive index for density and Hartung's empirical formula[18] for hydrocarbons is used to correct the density for the Hertzian pressure used in the test.

Film thickness data are generated by determining the entraining velocity (rolling velocity) corresponding to each fringe color. Film thickness tests were conducted at nominal temperatures of 23, 40, 70 and 100°C. Tests were limited to 23 and 40°C, for the fluids containing water, due to the loss of water by vaporization at higher temperatures. The tests were conducted with a recirculating fluid supply in a heated chamber. A trailing thermocouple was used to measure the ball temperature. A computer controlled peristaltic pump was used to recirculate the fluid.

The test results are graphically displayed using three types of plots: (1) EHD film thickness (h_o) vs. entraining velocity (U_e, m/s), (2) dimensionless film thickness parameter (h_o/R) vs. dimensionless speed parameter ($\eta_o U_e / E'R$) and (3) pressure-viscosity coefficient (Alpha, GPa^{-1}) vs. temperature (°C).

B. Performance Map Construction

Lubrication performance is reflected in both physical and chemical attributes of the hydraulic fluid. Designers, in general, use physical properties to predict in-use performance. Yet, most conventional bench tests, focus on the chemical and metallurgical properties of the contact pair under specific operating conditions. The tribological conditions being studied are frequently unrelated to those actually encountered in a specific pump wear regime. A single wear or scuffing test number does not adequately model the multi-dimensional character of a lubricant. Multi-dimensional characterization of a hydraulic fluid should also include such attributes as "antiwear," EHD film-forming ability, and traction coefficient. Non-pitting performance is also an attribute of interest. The synergism between the viscous film forming ability and the boundary film forming ability of a fluid determines its quality. The attributes of both are necessary for a proper performance determination.

A more comprehensive approach to lubricant evaluation is to model or "map" the performance of a contact system over a range of entraining and sliding velocities. Performance maps for two model hydraulic fluids (Fluids 1 and 2) and two reference oils (Fluids 3 and 4) are shown in Figures 6-9, respectively.[8] These figures show that Fluid 1 has substantially greater mixed-film or boundary lubricating performance than either Fluid 2 or 3, and especially Fluid 4. The viscous film-forming attributes of each fluid are judged by the entraining velocity necessary to bring the contact into the EHD region. While Fluid 4 has limited boundary lubricating properties, its EHD film-forming ability is superior to the other fluids. Fluids with good viscous film forming properties may sometimes mask their chemical or boundary lubricating deficiencies.

The relative size of the mixed-film lubrication region reflects the synergism between the viscous and boundary film lubricating ability of the fluid. Fluid 1 shows good performance in this regard. The scuffing or severe wear boundary reflects the fluid and material attributes with respect to their ability to prevent a sudden loss of surface integrity, or a transition into a severe wear mode. The location of this boundary does not necessarily reflect good wear resistance, since a sacrificial wear process can prevent the onset of a sudden scuffing event.

To further understand the physical meaning of these transitions, it is first necessary to understand the tribological concepts depicted by these maps. Performance maps are generated in terms of rolling (R) and sliding (S) velocity vectors. The generation of an EHD film is primarily a function of the entraining or rolling velocity (R) in the inlet region of the Hertzian contact region. In this region, the lubricating film generation, is primarily a function of the physical properties of the fluid (viscosity and pressure-viscosity coefficient). The sliding vector (S) determines the shear strain within the high-pressure Hertzian contact region. This region is important with respect to heat generation, surface film formation, wear and scuffing within the tribo-contact. The magnitude of the sliding velocity in the Hertzian region, along with the degree of surface interaction, invoke the chemical properties of the fluid, e.g., adsorbed films, chemical reaction films, tribochemical reactions and thermal/oxidative stability.

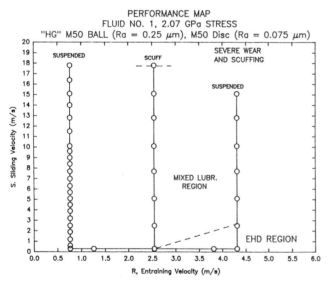

Figure 6 - Performance map for hydraulic Fluid No. 1

To create a performance map, a series of 10-minute tests are conducted over a wide range of rolling and sliding velocities. The M50 steel specimens are run with a contact stress of 2.07 GPa.[8] The ball specimen has a "hard grind" surface finish of Ra = 0.25 μm. The disk specimen is ground to a finish of Ra = 0.076 μm. Although M50 steel specimens were used here, other materials can be used for specific hardware simulation.

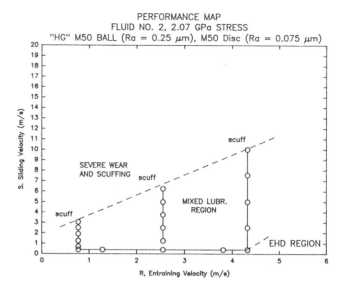

Figure 7 - Performance map for hydraulic Fluid No. 2

Lubrication and failure regimes are identified by the presence or absence of wear on the specimens, as well as transitions into a severe wear or scuffing failure. Specimen temperatures and the traction (friction) between the specimens are measured for each test.

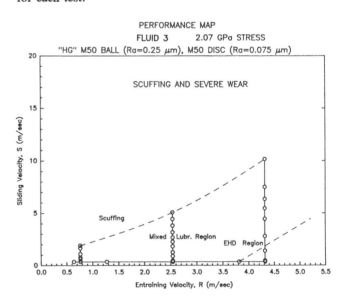

Figure 8 - Performance map for reference oil No. 3.

To identify the transition between "full-film" EHD lubrication and mixed-film lubrication, the running tracks on the specimens are microscopically examined and photographed as the sliding velocity is increased. The flexibility of the test machine allows the sliding velocity to vary while the rolling velocity remains constant. The transition point from full-film EHD lubrication is defined as the operating condition where local surface polishing on an asperity scale is observed. As the sliding velocity increases, the "hard grind" finish (Ra =

0.25 μm) on the ball becomes smoother. The polishing of surface features can be considered as the initiation of a run-in process, at least from a topographical viewpoint. The formation (and removal) of surface films, are expected to become more predominant as "S" increases. Also, there is greater heat generation which results in a temperature rise as the sliding velocity increases. The transition from mixed-film lubrication to scuffing, or in some cases to severe wear, is determined by running each fluid at a constant velocity (R = 80, 250 and 430 cm/s) with increasing sliding velocities until a scuffing event is observed. A scuffing transition is characterized by a sudden increase in traction coefficient in response to a complete breakdown of the lubricating film and gross smearing of the contact surfaces. Figure 10 shows a photomicrograph of the running track on the ball specimen following a scuffing event.

The traction derived from an EHD fluid film is a function of the limiting shear strength of the pseudo-solid fluid in the Hertzian contact region for highly loaded contacts. The traction coefficient decreases with increasing temperature. The decrease in traction is usually a linear function with respect to temperature. Monitoring the traction coefficient as a function of increasing sliding velocity reflects the nature of the sheared material within the contact region. In addition to bulk hydraulic fluid, the sheared material is likely to be composed of wear particles and breakdown products. The traction coefficient is sensitive to surface roughness, especially with high sliding velocities.

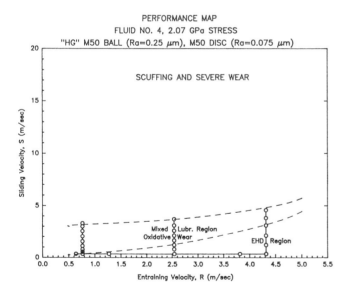

Figure 9 - Performance map for reference oil No. 4.

For most normal operating conditions, bulk fluid traction plays a predominant role in the overall traction at the contact. The contribution of "asperity" friction is only significant under severe operating conditions. This is shown in Figure 11, which is a plot of the traction coefficient vs. rolling velocity (R) for each fluid at a constant sliding velocity (S = 0.36 m/s). According to Figures 5-8, a rolling velocity over 4 m/s puts each fluid in the full-film EHD region. As the rolling

Figure 10 - Photomicrograph of a ball specimen following a scuffing event.

velocity decreases, the contact enters the mixed-film lubrication region. The four fluids have considerably different traction behavior. Fluid 1 has only 30% the traction coefficient of Fluid 4, even with its operation deep into the

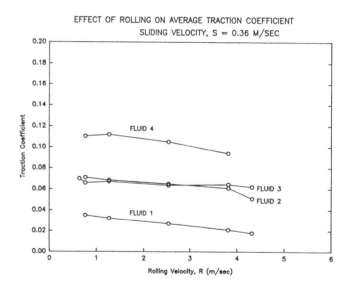

Figure 11 - Effect of varying rolling speed at a constant sliding velocity.

mixed-film lubrication region. Traction behavior is an important design consideration in view of its direct link to heat generation and its influence on the tangential shear stress at the contact surface. However, traction coefficients are not usually reported with fluid properties.

These data show that the determination of the location of the boundaries for lubrication and failure regions provide only a starting point for a more comprehensive performance characterization of the fluid. The fluid can be further characterized for pressure-viscosity properties that impart

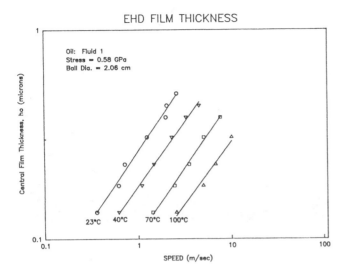

Figure 12a - Effect of rolling speed and temperature on film thickness for fluid No. 1.

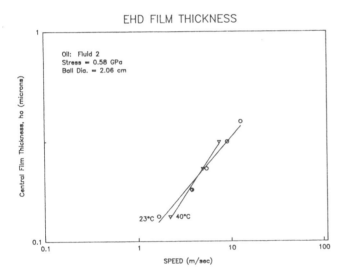

Figure 12b - Effect of rolling speed and temperature on film thickness for fluid No. 2.

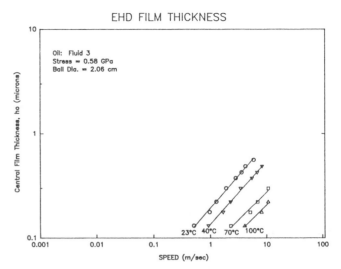

Figure 12c - Effect of rolling speed and temperature on film thickness for reference oil No. 3.

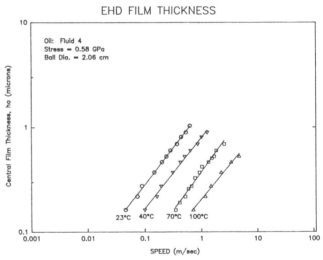

Figure 12d - Effect of rolling speed and temperature on film thickness for reference oil No. 4.

EHD film-forming ability. Testing along specific pathways through the mixed-film lubrication region can provide wear and micro-pitting characterization.

C. EHD Film Thickness Results

In the above discussion, it was shown that performance maps identify the regions of full-film EHD lubrication, mixed film lubrication and scuffing. The location or size of the EHD region reflect the viscous film-forming properties of the fluid. These are usually divided into viscosity-temperature and viscosity-pressure properties. The viscosity (η_o) and pressure-viscosity coefficient (α) are required for EHD film thickness calculations. While viscosity-temperature properties are

generally available, the pressure-viscosity coefficient is frequently missing. As part of a comprehensive characterization of fluid performance with the WAM3 machine, the pressure-viscosity coefficient can be determined from the measurement of EHD film thickness.

The viscosity, specific gravity and refractive index data for the test fluids are provided in Tables 1 and 2. Independently measured pressure-viscosity coefficients are given in Table 3.[15] Experimentally measured "effective" pressure-viscosity coefficients are shown in Table 4.

The film thickness data in Figures 12a-d show the effect of rolling speed and temperature on film thickness for each fluid. The log-log plots in these figures show that the relationship between film thickness and speed is exponential. From EHD theory, film thickness varies with entraining velocity to the power of 0.67 as seen in EQ(1).

$$h_o = 1.9 \frac{(\eta_o U_e)^{0.67} \alpha^{0.53} R^{0.397}}{E^{0.073} \, w^{0.067}} \qquad (1)$$

where
h_o = Film thickness, center of contact.
R = Combined radius of curvature.
η_o = Viscosity @ atm. press. & test temp.
α = Pressure-viscosity coefficient.
U_e = Entraining vel., $U_e = 1/2(U_1 + U_2)$.
E' = Combined elastic modulus.

Film thickness decreases with increasing temperature because the viscosity (η_o) and the pressure-viscosity coefficient (α) decrease with temperature. The slope of film thickness versus entraining velocity for Fluid No. 2, which contains water, is greater at 40°C than it is for 23°C. This may be due to water evaporation throughout the test, especially at higher temperatures (>70°C). Because of water evaporation, the pressure-viscosity determination for this fluid data is limited to the two temperatures of 23°C and 40°C.

The four test fluids are plotted in dimensionless form for the two temperatures in Figures 13a,b. The only EHD lubricant characteristic missing from the two dimensionless parameters is the pressure-viscosity coefficient (α) and load (w). The film thickness measurements were conducted under the same load. If the α-value for each fluid measured were the same, the film thickness data would all fall on a single line. Higher values of (h_o/R) for the same ($\eta_o U_e/E'R$) reflect higher pressure-viscosity coefficients.

TABLE 1

Viscosities and Specific Gravity of Experimental Hydraulic Fluids

Fluid No.	Temp. (°C)	Viscosity (Cst)	Specific Gravity
1	-17.8	1202	
	0	325	
	25	82.0	
	37.8	82.0	0.9088
2	0	340	
	20		1.089
	40	46	
	65	22	
3	37.8	21.4	0.998
	98.9	4.5	

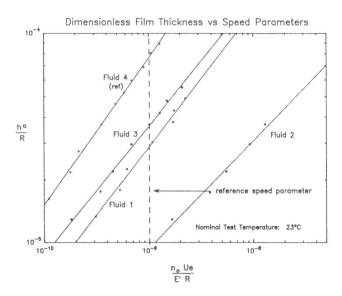

Figure 13a - Dimensionless film thickness vs. speed for each fluid at 23°C.

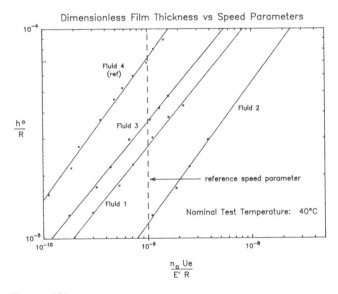

Figure 13b - Dimensionless film thickness vs. speed for each fluid at 40°C.

Calculation of Pressure-Viscosity Coefficient - For each test fluid, the film thickness data was plotted in dimensionless form. The dimensionless "film thickness parameter" (h_o/R), where:

h_o = film thickness in the center of the contact,
R = combined radius of curvature (0.3968 in.),

and the speed parameter EQ(2), which is defined as:

$$\frac{\eta_o U_e}{E' R} \qquad (2)$$

47

where: η_0 = viscosity at atmospheric pressure and test temperature,

   U_e = entraining velocity, $1/2(U_1 + U_2)$,

   E' = combined elastic modulus of specimen materials,

   R = combined radius of curvature (0.3968 in.)

The only EHD lubricant characteristic missing between these two parameters is the pressure-viscosity coefficient (α). If the α - value for each fluid measured were the same, the film thickness data would all fall on a single line. Test results that depart from the reference fluid data, reflect variations in α.

Figure 13a,b illustrates the test data plotted in the dimensionless form using speed and film thickness parameters, which includes all of the parameters that govern EHD film generation, except for the pressure-viscosity coefficient (and load which is constant). This figure shows that the data for fluids 1-4 all fall on significantly different lines, therefore, the pressure-viscosity coefficients are very different.

Using Fluid No. 4 as the reference fluid (since the pressure-viscosity of this fluid is well characterized), it is possible to calculate the "relative" pressure-viscosity coefficient from the Hamrock-Dowson Equation, EQ(3):

$$\alpha = \alpha_{ref\,oil} \left[\frac{h_o/R}{(h_o/R)_{ref\,oil}} \right]^{1.887} \qquad (3)$$

The above equation assumes that h_o is proportional to $\alpha^{0.53}$ according EHD theory. The relative pressure-viscosity coefficients for the test fluids were determined at a selected speed parameter of 1×10^{-9}. This value was selected because it has corresponding film thickness data for all the test fluids. The calculated α-values are shown in Table 5.

These data are very interesting since they show that the optically measured pressure-viscosity coefficient (Table 4) is always lower than the value obtained by capillary viscometry (Table 3), and by approximately the same amount. Of the various potential reasons for this, perhaps one of the most significant is the temperature-viscosity effect. These data suggest that the pressure-viscosity and temperature-viscosity coefficients do not vary independently of each other. This would be reasonable since the traction coefficients for these lubricants are different, the contact temperatures are also probably different. If the contact temperature varies between the various fluids, the actual temperature of measurement of the film temperature may vary somewhat. Therefore, these results reflect an "effective" pressure-viscosity coefficient actually experienced by the fluid in the process of film generation. However, more work is required to definitively explain this behavior. Therefore, the variation of pressure-viscosity coefficient with temperature is shown Figure 14. Interestingly, this data shows that the pressure-viscosity coefficient of Fluid No. 4 is more sensitive to temperature

than Fluid No. 3. It is generally known that fluids with sensitive viscosity-temperature properties also have sensitive viscosity pressure characteristics. The slight increase in viscosity-pressure characteristics of the water containing Fluid No. 2 is attributed to water evaporation.

TABLE 2

Refractive Indices of Experimental Hydraulic Fluids

Fluid No.	Pressure (psi)	Refractive Index
1	Atm.	1.4690
	92,000	1.5799
2	Atm.	1.4145
	92,000	1.5091
3	Atm.	1.45
	92,000	1.559
4	Atm.	1.2985
	92,000	1.362

TABLE 3

Measured Pressure-Viscosity Coefficients of Experimental Hydraulic Fluids

Fluid No.	Temp (°C)	Pressure-Viscosity Coeff. (GPa⁻¹)
1	65	11.5
2	65	2.97
3	65	10.0
4	65	37.0

CONCLUSION

An important, and often little understood, function of hydraulic fluids is to lubricate the hydraulic pump where either sliding, e.g. vane on ring in a vane pump, or rolling, e.g. rolling element bearings, contact velocities and related wear may result. There are various methods of studying these effects which include wear tests in hydraulic pumps, the use of custom designed bench tests or the use of a test methodology which permits the study of potential wear in its most fundamental form by modeling traction and scuffing as a function of rolling and sliding speed and load.

The use of performance maps to identify the boundary regions of various wear regimes along with fluid traction behavior within these regions has been demonstrated using a new and comprehensive test approach. The test approach includes a

TABLE 4

PRESSURE-VISCOSITY COEFFICIENTS

TEST NO.	OIL	TEST TEMP ($^\circ$C)	H_o/R Ref. Oil	H_o/R Test Oil	ALPHA Ref. Oil (GPa)$^{-1}$	ALPHA Test Oil (GPa)$^{-1}$
194	Fluid No. 1	24	7.30E-05	2.90E-05	4.78E-08	8.37E-09
195		38	7.00E-05	2.90E-05	4.13E-08	7.83E-09
196		68	6.50E-05	2.90E-05	3.57E-08	7.78E-09
197		92	6.40E-05	3.00E-05	3.22E-08	7.71E-09
198	Fluid No. 2	25	7.30E-05	1.03E-05	4.78E-08	1.19E-09
199		41	7.00E-05	1.30E-05	4.13E-08	1.72E-09
204	Fluid No. 3	25	7.30E-05	3.80E-05	4.78E-08	1.39E-08
205		41	7.00E-05	3.70E-05	4.13E-08	1.24E-08
206		70	6.50E-05	3.20E-05	3.57E-08	9.37E-09
207		90	6.40E-05	3.30E-05	3.22E-08	9.23E-09
220	Fluid No. 4	30	7.30E-05	7.30E-05	4.78E-08	4.78E-08
221		41	7.00E-05	7.00E-05	4.13E-08	4.13E-08
222		72	6.50E-05	6.90E-05	3.57E-08	4.00E-08
223		96	6.40E-05	6.40E-05	3.22E-08	3.22E-08

TABLE 5

Pressure-Viscosity Coefficients at 65°C Obtained By Capillary Viscometry and EHD Film Measurements (GPa⁻¹)

Fluid No.	Capillary Viscometer	Optically Measured from EHD Contact
1	11.5	7.8 (68°C)
2	2.97	1.8 (61°C)
3	-	11
4	37	37 (ref. fluid)

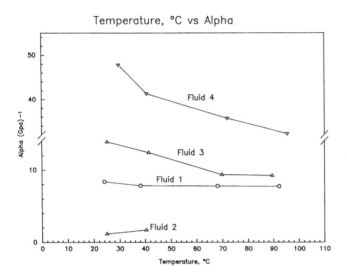

Figure 14 - Pressure-viscosity coefficient vs. temperature.

capability to measure the oil film thickness within the contact region. Film thickness measurements conducted over a range of temperatures can be used to calculate an "effective pressure-viscosity coefficient." Although the "effective pressure-viscosity coefficient" is consistently less than that measured by capillary viscometry, it is expected to reflect the dynamic conditions invoked by the film generating process within the EHD contact.

This work demonstrates the value of examining hydraulic lubrication and wear from a fundamental, first-principles approach. This approach does not require numerous specialized tests, nor does it require the frequent use of relatively expensive, large volume hydraulic pump tests. With the availability of the appropriate information, it is possible to use such fundamental lubrication calculations in the design phase of hydraulic pump development and subsequent fluid qualification. This will be the subject of a future report.

REFERENCES

1. K. Mizuhara and Y. Tsuya, <u>Proc. of the JSLE Int. Tribology Conf.</u>, (1985), Tokyo, Japan, p. 853-858.

2. R.K. Tessmann and I.T. Hong, <u>SAE Technical Paper Series</u>, Paper Number 932438 (1993).

3. G. Silva, <u>SAE Trans.</u>, **99**, (1990), 635-652.

4. R. Stribeck, <u>Zeit V.D.</u> 1, **46**, (1902).

5. G. Jacobs, W. Backe, C. Busch and R. Kett - IFAS - Aachen, Germany, "A Survey on Actual Research Work in the Field of Fluid Power", Paper to be presented at the 1995 STLE National Meeting, Chicago, IL

6. L.D. Wedeven, "Thin Film Lubrication and Tribological Simulation", in <u>Adv. in Eng. Tribology</u>, STLE SP-31, April, 1991, p. 164-184.

7. U.S. Patent Application Number: 07/963,456.

8. L.D. Wedeven, G.E. Totten, R.J. Bishop Jr., <u>SAE Technical Paper Series</u>, Paper Number 941752, (1994).

9. G.T.Y. Wan, P. Kenny and H.A. Spikes, <u>Tribology Intl.</u>, **17**, (1984), p. 309-315.

10. H.A. Spikes, <u>J. Synth. Lubr.</u>, **4**, p. 115-135.

11. G.T.Y. Wan, <u>ASLE Trans.</u>, **27**, (1984), p. 366-372.

12. V.P. Veresnyak, L.V. Zaretskaya, T.V. Imerlishvili, V. Kel ' bas, N.V. Lukashvili, L.O. Sedova and V.M. Ryaboshapka, <u>Trenie Iznos</u>, **10**, (1989), p. 919-27.

13. V.P. Veresnyak, L.V. Zaretskaya, T.V. Imerlishvili, V. Kel ' bas, N.V. Lukashvili, L.O. Sedova, V.M. Ryaboshapka, V. Sh. Shvartsman and V. Kh. Shoikhet, <u>J. Frict. Wear</u>, **10**, p. 120-126.

14. V.P. Veresnyak, L.V. Zaretskaya, T.V. Imerlishvili, V. Kel ' bas, N.V. Lukashvili, L.O. Sedova, V.M. Ryaboshapka, V. Sh. Shvartsman and V. Kh. Shoikhet, <u>Trenie Iznos</u>, **12**, p. 144-153.

15. S. Bair and W.O. Winer, <u>Tribol. Trans.</u>, **31**, (1987), p. 317.

16. H. Ueno and K. Tanaka, <u>Junkatsu</u> (J. Jap. Soc. Lubr. Eng.), **33**, (1988), p.425-430.

17. S. Glasstone, **Textbook of Physical Chemistry**, Macmillan and Co. Ltd., London, 1948.

18. P.S.Y. Chu and A. Cameron, "Compressibility and Thermal Expansion of Oils", <u>J. Inst. Petrol</u>, **49**, (1963), p. 140-5.

Hydraulic Pump Testing Procedures to Evaluate Lubrication Performance of Hydraulic Fluids

G. E. Totten and R. J. Bishop, Jr.
Union Carbide Corp.

ABSTRACT

Although the selection and role of hydraulic fluids as energy transfer agents is relatively well understood, there is no consensus on the appropriate procedures to evaluate lubrication properties on a laboratory scale. Because the use of bench tests such as the Shell 4-ball has traditionally produced poor pump wear correlations, it has been necessary to develop various hydraulic pump tests for this purpose. Since hydraulic fluid lubrication is being modeled, it is necessary to view these hydraulic pump tests as *tribological tests*. The objective of this paper is to provide an overview of various vane, piston and gear pump tests that have been reported as tribological tests.

INTRODUCTION

Fluids play two vital roles in hydraulic pump operation. Perhaps the best known and understood is their role in energy transfer. However, their role as lubricants is more poorly understood. This problem is compounded because there are very few standardized methods of evaluating and reporting the relative ability of a hydraulic fluid to adequately lubricate the various critical components in a hydraulic system.

Silva published a thorough review of the wear mechanisms in hydraulic pump operation.[1] The role of cavitation, adhesion, corrosion and abrasion wear was described. Also discussed was the role of fluid viscosity and the speed and loading at the wear contact as modeled by the classic Stribeck curve illustrated in Figure 1. However, discriminating methods of experimental modeling of pump wear was not discussed.

There are numerous references to the use of hydraulic pump tests to evaluate component durability[2,3] or to evaluate some aspect of component design features on either the efficiency or mechanism of energy transfer. However, there are fewer references on the development and use of hydraulic pumps as "tribological tests" to evaluate fluid wear.

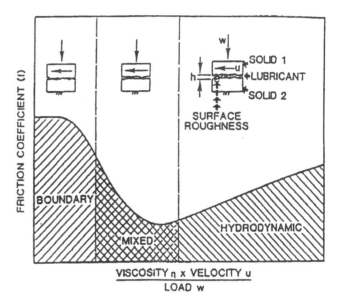

Figure 1 - Illustration of the effect of wear contact loading and speed and fluid viscosity on wear (Stribeck curve).

There have been some references describing the impact of fundamental lubrication properties of hydraulic fluids, such as film thickness, on pump wear using the hydraulic pump as a tribological test.[4] However, there are a number of problems with such testing procedures which include: cost of pumps, energy and components, relatively long testing times, relatively poor manufacturing precision of some of the components for use in reproducible tribological testing, volumes of fluid required and subsequent disposal. Therefore, there has been a longstanding effort to develop "bench test" alternatives to evaluate hydraulic fluid lubricity.[5]

Examples of bench tests include: Shell 4-ball, Falex pin-on-V-block, and many others. Bench tests have a number of advantages relative to pump tests including: relatively low fluid volumes are required, numerous material variations and wear contact configurations are possible, high-quality and reproducible test components are readily available and substantially lower cost of testing.

Although it would be desirable to conduct bench tests as alternatives to hydraulic pump testing, with few exceptions these tests provide poor correlation with pump lubrication. For example, Renard and Dalibert found no correlation between either the Shell 4-ball "wear" (ASTM D-2596-T) or "extreme pressure" (ASTM D-2783) and the wear results obtained with a Sperry-Vickers V-104 vane pump test (ASTM D-2882).[6] Knight reported that: "...Regrettably the data from these tests not only failed to give quantitative correlation with data determined from machines, but even the ranking between different fluids failed to agree. ... ."[7] More recently, Lapotko, et.al. reported that 4-ball test results "... may deviate considerably from the data obtained in actual service..."[8]

Recently, Tessmann and Hong[9] and Perez, et.al.[10] have reported testing modifications of the Falex pin-on-V-block and Shell 4-ball tests respectively, that provide acceptable correlation with vane and ring wear in a Sperry-Vickers V-104 and 35VQ pumps.

Mizuhara and Tsuya extensively studied the wear correlation of a broad range of hydraulic fluids evaluated in a piston, vane and gear pump with a block-on-ring (Timken) test conducted according to ASTM D-2741-68.[11] From these studies, they concluded:

● "Load-carrying capacity" has nothing to do with hydraulic pump antiwear performance. An "antiwear test" must be used.

● Accelerated tests usually give the wrong results.

● Material pairs for the test pieces must be the same as the wear contact of interest in the hydraulic pump.

● It is necessary to evaluate hydraulic fluids under a wide range of test conditions. A single set of test conditions is usually inadequate.

One method of evaluating a hydraulic fluid under varying lubrication conditions is to construct a performance map such as that illustrated in Figure 2 to determine the mechanistic transitions between hydrodynamic, elastohydrodynamic (EHD), mixed EHD and boundary lubrication as a function of wear contact loading and rolling and sliding speed.[12] The anti-friction characteristics within each transition can then be evaluated as shown in Figure 3. Of course, as Mizuhara and Tsuya showed, this work should be performed with the appropriate wear contact material pairs.[11]

Although standardized bench tests have been generally shown to be unreliable, it is possible to modify these tests to achieve limited correlation with lubrication performance in a pump. Alternatively, pump lubrication can be examined fundamentally by modeling pump lubrication using a range of test conditions and appropriate material pairs. However, irregardless of what test is performed, it is always necessary to first calibrate the bench test to hydraulic pump wear and then to validate bench test conclusions by pump testing.

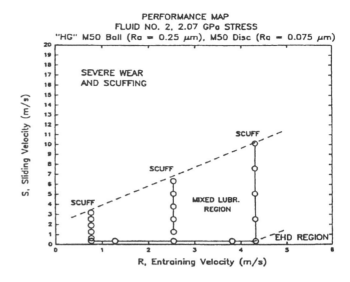

Figure 2 - Illustration of the mechanistic transitions of lubrication as a function of rolling and sliding speed.

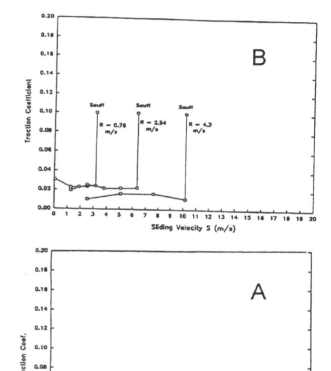

Figure 3 - Traction coefficient variation with varying rolling (A) and sliding (B) speeds.

Therefore, it is still critically necessary to evaluate hydraulic fluid lubrication performance by pump testing. The objective of this paper is to provide an overview of various pump testing procedures that have been used to study hydraulic fluid lubrication. This discussion will include vane, piston and gear pump testing procedures and also a summary of evaluation criteria.

DISCUSSION

A. Vane Pump Testing

The output pressure of a vane pump is directed to the back of the vanes which holds them against the ring. The leading edge of the vane forms a <u>line contact</u> with the ring and the rotation of the vane against the ring generates a <u>sliding motion</u>. This is the tribological condition being modeled by a vane pump test.

Hemeon reported the application of a "yardstick formula" to quantitatively analyze and report vane pump wear. The critical part of the analysis was weight loss of the ring, since dimensional changes due to wear will significantly affect volumetric efficiency and leads to noise and pulsations.[13] The yardstick formula is represented by EQ(1):

$$Y = K \times 1.482 \times mep \times gpm \qquad (1)$$

where: Y = Duty load on the pump in BTU/hr.
X = A constant to correct for air entrainment, degraded or contaminated oil and fluid turbulence.

Hemeon reported that although K may be as high as 1.4, he typically used a value of 1.03. The conversion constant 1.482 permits the use of pressures in psi and weight loss in grams.

Interestingly, in order to obtain reproducible and reliable weight loss data, it was reported that the ring had to be washed and baked at 200°F for 24 hours because the porous metal adsorbed solvent and gave incorrect weight data.

Shrey used the Sperry-Vickers V-104E vane pump test (see Figure 4) to evaluate the antiwear properties of both aqueous and non-aqueous fluids using a hydraulic circuit similar to that shown in Figure 5.[5] Most of Shrey's tests were conducted at 1000 psi, 1000 hours, 1.9 gal/min, 1120 rpm. The total weight loss of the vanes and ring were determined and reported.

A variation of this test, where the pressure of the pump cycled 20 seconds on and then 20 seconds off while the relief valve was set at 1000 psi to set up a "shock running load" from 0-1000 psi every 20 seconds. No other comparisons or conclusions were drawn other than to state that no damage to the moving parts was observed.

Bosch (Racine Fluid Power) also utilizes a cycled pressure vane pump test. The pressure-time sequence and test circuit is illustrated in Figure 6.[18] Based on their research, this is a more representative test since it better incorporates pressure

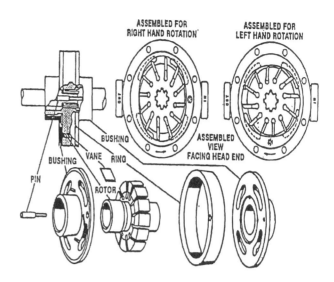

Figure 4 - Illustration of the Sperry-Vickers V-104 vane pump.

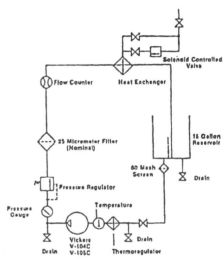

Figure 5 - Test circuit for Vickers V-104 vane pump tests.

spikes that will invariably be experienced by the system during circuit activation and deactivation. At the conclusion of the test, the weight loss of the ring, vanes, port and cover plate, and body and cover bearings were measured, the ring and bearings were inspected for unusual wear patterns and for evidence of corrosion, rusting and pitting.

Reiland used essentially the same constant pressure test circuit used by Shrey above to evaluate antiwear oils.[14] However, another variant of this now-classic test was evaluated. In this case, the V-104 vane pump was run for 100 hours at 2000 psi (the pump is only recommended for 1000 psi) and 1200 rpm. These are the test conditions currently incorporated in ASTM D-2882. The conclusion was that "the more severe the test, the better its ability to discriminate between oils."[14] However, these running conditions resulted in frequent breakage of the rotor (which still occurs today[29]).

The relative lubricity of a hydraulic oil was determined by visual comparison of wear curves that were constructed by plotting total weight loss of the vanes + ring versus total running time as shown in Figure 7.

$$P_a = 2000 \left[\frac{A_b}{A_a} \right] \qquad (2)$$

where:

P_a = unit pressure on the leading edge of the vane,

P_b = unit pressure on the back of the vane,

A_a = area of the leading edge of the vane,

A_b = area of the back of the vane.

For 2000 psi pump operating pressure, EQ(2)

becomes:

$$P_a = P_b \left[\frac{A_b}{A_a} \right] \qquad (3)$$

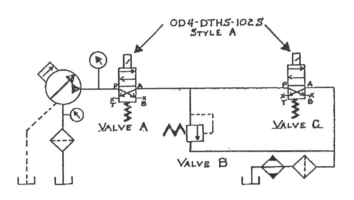

	0 psi	**100 psi**	**600 psi**
Valve A	ON	OFF	ON
Valve C	ON	ON	OFF
	10 sec	10 sec	10 sec
		← 1 cycle →	

Figure 6 - Test circuit for Racine cycled-pressure vane pump test utilizing a 7.5 hp electric motor, SV-10 vane pump and a 20 gallon reservoir[18].

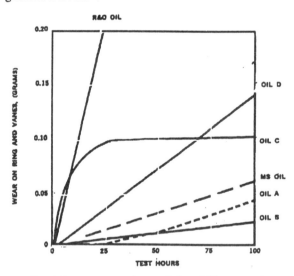

Figure 7 - Wear curves obtained for different antiwear oils using the Vickers V-104 vane pump test.

Alternatively, it was also proposed that antiwear properties of a hydraulic oil is also reflected by the "unit compressive pressure" which is determined by photomicrographing the leading edge of the vanes after the test, measuring the wear scar and solving EQ(2):

Weaver used a radiotracer technique to measure vane pump wear.[15] Either the vanes or the ring were irradiated and wear was measured by measuring the gamma radiation of the radioactive particles. While interesting, this procedure is of little practical value as an industry-wide wear test.

The Vickers V-104 continues to be the commonly utilized hydraulic fluid pump test.[29] There are at least three national standards based on the use of this pump; ASTM D-2882, DIN 51389, and BS 5096 (IP 281/77). A comparison of the test conditions is provided in Table 1. The total weight loss of the vanes + ring at the conclusion of the test is the quantitative value of wear.

Currently an effort is underway within ASTM (D.02 N.07 committee) to replace the V-104 pump with a more current design; the Vickers 20VQ vane pump. Some differences between the pumps are as follows:

	V-104	20VQ
Vane Load (lbs/1000 psi)	47	25
Max. Pressure (psi)	1000	3000
Max. Speed (rpm)	1500	2700
End Plates	Cast Bronze	Sintered Bronze

Initial round robin studies involving the simple replacement of the V-104 vane pump in the ASTM D-2882 procedure gave mixed results. One shortcoming, is that while the correlation of the results is generally the same, the agreement is relatively poor. Also, previously unreported test results have shown that while there seems to be a correlation between results obtained by the new pumps, absolute agreement (as expected) is poor. Results for some fluids when tested under the same conditions as shown in Table 2.

TABLE 1

Comparison of Sperry-Vickers V-104 Vane Pump Testing Procedures

Test Parameter	ASTM D-2882	DIN 51389	BS 5096 IP281/77
Pressure	14 MPa 2000 psi	10 MPa 1500 psi	14 MPa[2] 2088 psi 11 MPa[3] 1540 psi
Rev/Min	1200	1500	1500
Time (Hrs)	100	250	250
Fluid Vol. (liters)	56.8		55-70
Fluid Temp	150° F	[1]	[1]

1. The fluid temperature is selected to give 46 Cst at the test temperature.

2. The pump is run at 140 bar (2088 psi) for mineral oil type fluids.

3. The pump is run at 105 bar (1540 psi) for HFA, HFB and HFC hydraulic fluids.

Lapotko, et.al. have reported an alternative vane pump test designated as the "MP-1 Test".[8,16] A schematic of the MP-1 vane pump is shown in Figure 8.

Although the MP-1 may be run at pressures up to 10 Mpa (1450 psi), the reported test pressure is 7 Mpa with a total fluid volume of 0.7 liters. In addition to lower volume, the MP-1 test is conducted for only 50 hours (and in some cases for only 10 hours). The wear rate is based on the weight loss of the vanes only after the test is completed. This test comes as close to a "bench hydraulic pump test" as has been reported to date.

One of the deficiencies of all of the previously described vane pump tests is that they are either conducted at relatively low pressure or rpm, compared to vane pumps in industrial use. Although not a national standard, the Vickers 35VQ pump test

TABLE 2

Comparison of the Wear Rates Obtained With Various Water-Glycol Hydraulic Fluids Using the ASTM D-2882 Test Conditions and the Vickers V-104 and 20VQ Vane Pumps(1)

FLUID	WEAR RATE (mg/hr)[2]	
	20VQ	V-104
1	0.45	4.1
2	0.11	0.37

1. The test conditions were 2000 psi, 1200 rpm and 100 hours.

2. The wear rate is obtained by dividing the total weight loss of the vanes and ring by 100 hours to give wear rate (mg/hr).

is often acknowledged to provide a more rigorous, and in some cases, a more realistic accelerated industrial pump wear test (20.7 MPa, 3000 psi/2400 rpm).[17] A schematic of this test is provided in Figure 9.

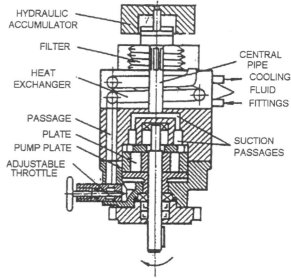

Figure 8 - A schematic illustration of the Russian MP-1 vane pump.

The test criterion is for a maximum weight loss of the vanes for each the test cartridges must not exceed 15 mg and the total weight loss of the ring for each cartridge must not exceed 75 mg and three cartridges are to be used. If the weight loss of any of the cartridges exceeds these values, two more tests with new cartridges must be run.

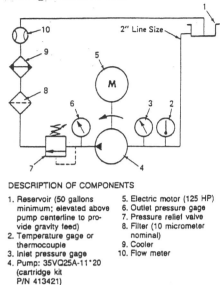

Figure 9 - Schematic of the Vickers 35VQ test stand.

Hagglund-Dennison have also developed a two-part recommended pump testing protocol (HF0). Although this test is used widely in the hydraulics industry, it is not a national standard test. The HF-0 procedure is composed of two parts, a vane pump test and a piston pump test. Only the vane pump test will be discussed at this point.

The test circuit for the Hagglunds-Dennison vane pump wear test is based on their T5D-042 vane pump, which is located in a "bootstrap" circuit with the Dennison series 46 piston pump. The T5D vane pump test circuit is shown in Figure 10. The vane pump is operated at 2400 rpm and 2500 psi and is driven by a 136 hp electric motor. At the conclusion of the test, the procedure requires that the following be recorded:

- Cam-rotor and cam-vane clearance,

- Vane-and-slot clearance for all vanes,

- Tracings of lip contours of vanes 1,4 and 8,

- Visual appearance of cam ring interior surface,

- Visual appearance of both plate surfaces that are against rotating pumping elements,

Hagglund-Dennison has also developed a SK-30320 vane pump test to evaluate vegetable oil durability. This test is also based on the Hagglunds-Dennison T5D-042 vane pump and is a 600 hour, cycled pressure test (300 hours under "dry" conditions and then 300 hours after the addition of 1% of water. The pressure cycles from 10 to 240 bars each second and the fluid temperature is 70°C.

Most of the references described above utilize a gravimetric

determination of wear by either measuring the weight loss of the vanes, ring or both. However, some hydraulic fluids

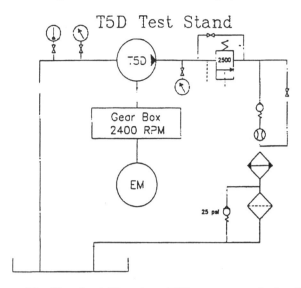

Figure 10 - Hagglunds-Dennison T5D vane pump hydraulic test circuit.

exhibit non-newtonian viscosity behavior leading to internal leakage and loss of power transmission efficiency. These problems can detected by experimental determination of pumping efficiency by flow rate measurement[19].

B. Piston Pump Testing

There are numerous wear surfaces in a piston pump. In addition to sliding wear, such as pistons in cylinders, there is mixed rolling and sliding which would occur with rolling element bearings, corrosion and cavitation wear which might occur on the swash plate, etc. The relative amount of wear that would occur would also be critically dependent on the material pairs of construction of the wear contacts. In view of the wide range of materials used for construction and designs of piston pumps, this is one reason why most of the "standard" pump tests developed to date have been vane pump tests.

The lubrication challenges in a piston pump are illustrated in the piston pump depicted in Figure 11. The objective of piston pump design is to minimize energy consumption while at the same time optimizing hydrodynamic lubrication to minimize wear and to internal leakage. The performance parameters are fluid flow, speed, torque, pressure, viscosity, and inlet pressure. To minimize friction and internal leakage, wear contact loading (pressure), speed and viscosity must be optimized as shown in Figure 1.[1] In a piston pump, the piston clearances may vary with "eccentricity" due to load and fluid viscosity. This may produce a change in the lubrication mechanism, e.g. hydrodynamic to boundary, resulting in increased wear and friction.

TABLE 3

500 Hour Cycled Pressure Sundstrand Pump Test Sequence

Duration[1] (sec)	Vickers (MPa)	Sundstrand (MPa)
130	1.17	21.37
325	0.72	17.24
60	2.07	31.03
85	0.72	17.24

1. The total time per cycle is 600 seconds.

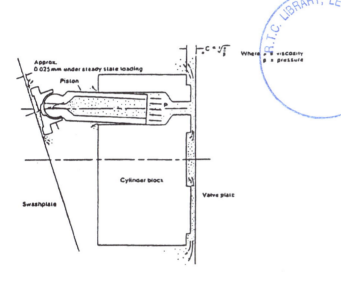

Figure 11 - Illustration of the clearances in a piston pump.

One piston pump test commonly used in the hydraulic fluid industry is the "Sundstrand Water Stability Test Procedure".[20] The test circuit containing a Sundstrand Series 22 axial piston pump is shown in Figure 12. The test conditions are:

Input Speed	3000 - 3200 rpm
Load Pressure	5000 psi
Charge Pressure	180 - 220 psi
Case Pressure	40 psi max.
Stroke	1/2 of full
Reservoir Temperature	150 +/- 10°F
Loop Temperature	180 +/- 10°F
Maximum Inlet Vacuum	5 Inches Hg

The objective of this test is to determine the effect of water

contamination on mineral oil hydraulic fluid performance. The duration of the test is 225 hours. In addition to disassembly and inspection for wear, corrosion and cavitation, the test criteria is "flow degradation". Flow degradation of 10% is considered a failure.

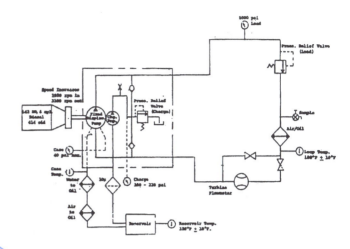

Figure 12 - Schematic of the Sundstrand Water Stability Test Circuit

A cyclic loading variation of this test was recently reported to evaluate the performance of a water-glycol hydraulic fluid under these relatively high loading conditions.[21,22] In this test, a Sundstrand swash plate, axial piston pump was tested at 3175 rpm which was driven by a Sundstrand Series 20 swash plate piston motor at 890 rpm. The load was supplied by a Vickers vane pump at 975 rpm. The test sequence is summarized in Table 3. The rotating components are visually inspected for wear.

The test circuit for the piston pump portion of the Hagglunds-Dennison HF-0 piston pump is illustrated in Figure 13. This test, which utilizes a Dennison P46 piston pump was developed to evaluate the multi-metal compatibility of a fluid and its corrosiveness against soft metals in a severe hydraulic environment. This test is conducted for 100 hours and then the pump components are visually inspected for wear, corrosion and cavitation at the conclusion of the test.

Another piston pump test has recently been proposed to ASTM Committee D.02 N.07 by The Rexroth Corporation.[23] The test circuit is provided in Figure 14 and is based on a Breuninghaus A4VS0 swash plate, axial piston pump. The objective of this test is to better discriminate and classify hydraulic and wear performance of a hydraulic fluid. It is proposed that this would be done by prescribing performance levels to be achieved. Establishment of these performance levels has not been developed as of this time.

Another less well known cycled pressure piston pump test is the Vickers AA 65560-1SC-4 piston pump test.[24] Previous work showed that if there is no cavitation resulting from inlet

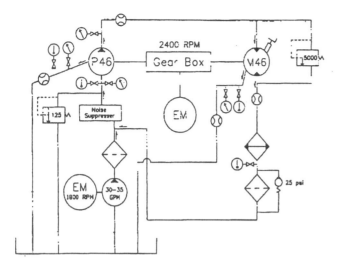

Figure 13 - Schematic of the Dennison P46 piston pump test portion of the Hagglunds-Dennison HF-0 protocol.

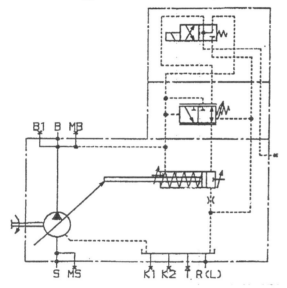

Figure 14 - Schematic of the proposed Rexroth Piston Pump Test.

starvation and if the hydraulic fluid is clean, then there are four principle modes of failures: 1.) spalling of the yoke and spindle bearing group, 2.) fatigue failure of the control piston assembly, 3.) rotating group assembly failure due to worn front and tail drive bearings and 4.) static and dynamic O-ring failures. In a sense, piston pump tests are excellent bearing and cavitation tests.

Janko used high-speed piston pump tests conducted for 1250 hours to supplement successful preliminary vane pump testing.[25] This test was conducted at constant 140 bar pressure for 1000 hours and then completed by cycling the pressure between 70 and 140 bars at 0.1 Hz. The pump was disassembled every 250 hours and visually inspected. The stressed components including the pistons, piston slippers, cylinder barrel, and reversing plate were inspected and

measured for wear. In addition, it was learned that the hydraulic fluids being studied caused such severe bearing wear that they had to be replaced every 250 - 500 hours. The same fluids produced severe wear on the drive shaft with spline ring and thrust pins. These results show the value of piston pump testing to significantly increase the wear stress of hydraulic fluids.

Edghill and Rubbery studied the correlation of laboratory testing with the type and frequency of field failures.[26] There analysis showed that the most critical areas for failure are:

For Strength:
    1. Piston Necks
    2. Shafts
    3. Body Kidney Ports

For Bearing Surfaces:
    1. Slipper/Cam Plate Interface
    2. Piston Skirt /Cylinder Bore
    3. Cam plate Trunion Liners

Some piston pumps are used in very broad range temperature environments (-46 - 204°C). There are two tests that have been reported for these applications. Hopkins and Benzing utilized the test circuit shown in Figure 15, which utilized a Manton Gaulin Model 500 HP-KL6-3PA, three-piston pump. The modifications used for this pump did not facilitate analysis of pump wear surfaces. Instead, the tests which were conducted at 3000 psi, 550°F for 100 hours were designed to evaluate fluid degradation, corrosion, lacquering and sludging tendencies.

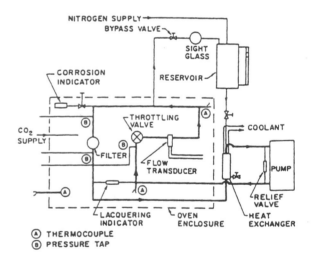

Figure 15 - Hopkins and Benzing high temperature modified Manton Gaulin piston pump test circuit.

Gschwender, et.al. used a test circuit utilizing a Vickers model PV3-075-15 pump to evaluate the wear of high temperature (122°C/250°F) poly(alpha olefin) based hydraulic fluids.[28] The test circuit is illustrated in Figure 16. The tests were conducted at 20.4 MPa (2960 psi) and 5400 with the throttle valve closed and 5000 psi with the throttle valve open.

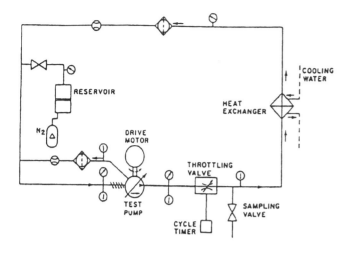

Figure 16 - Gschwender piston pump test stand.

3. Gear Pump Testing

In the above discussion regarding vane pumps, it was observed that the primary mode of wear, although certainly not the only one, was sliding wear of the vanes on the ring. For piston pump testing, failure analysis was more complex since there were numerous surfaces that must be inspected. Sliding with hydrodynamic wear was still the primary component. However, while sliding wear is still important in gear pumps, the wear mechanisms are even more variable. For example, in open gears, hydrodynamic, EHD, mixed EHD and boundary lubrication mechanisms may all occur simultaneously, depending on the position, and speed of the gear as seen in Figure 17.[30]

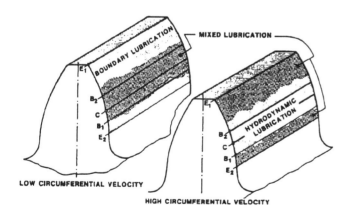

Figure 17 - Lubrication mechanisms on the gear tooth as a function of speed.

Frith and Scott performed a detailed theoretical analysis of gear pump wear. They noted that the primary areas of wear are the side plate near the suction port, the gear meshing zone,

the gear tips and casing, especially near the suction port.[31] In addition, the imbalance of pressure across the pump may cause the gear shaft to deflect toward the inlet creating a reduction in gear tip clearance at the inlet. Also, a hydrodynamic wedge between the gear ends and the side plate, in combination with the pressure behind the plate, tends to force the side plate against the gears in the inlet region as shown in Figure 18.

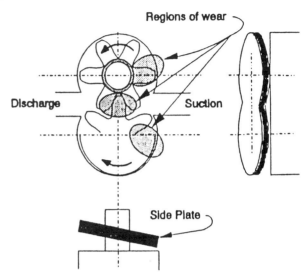

Figure 18 - Side plate action in a gear pump.

All of these conditions may affect the efficiency of gear pump operation. Interestingly, there have been relatively few reports of hydraulic fluid lubrication in a gear pump and there are no industry standard gear pump lubrication tests.

One study that has been reported was conducted by Knight who used the multiple gear pump test stand shown in Figure 19. Seven Hamworthy Hydraulics Ltd. Type PA 2113 gear pumps were run at 14.3 MPa (2075 psi). The test is run until pump failure, usually due to needle bearing fatigue. The condition of the roller bearings are monitored at least once every 24 hours.

In another study, a cycled load over 500,000 cycles from zero to the maximum rated pressure, speed and temperature for the pump and the fluid was recommended.[32] In addition to the cycled loading test, an endurance test (maximum pressure, speed and temperature for the fluid for 250 hours, "proof" test where the pump is operated under extreme conditions at relatively short periods of time, e.g. 5 hours and an initial run-in test. Toogood stated that although degradation in performance would occur under extreme conditions, it was not clear if the damage was greatest under constant pressure or cycled pressure conditions.[32]

Wanke conducted a study of the effects of fluid cleanliness with a multiple gear pump test stand shown in Figure 20.[33] Two test sequences were used. One was a cycled pressure test (1 cycle/10 seconds). The other sequence was an endurance test. As a result of this work, it was recommended that although flow was an adequate measure of the pump's

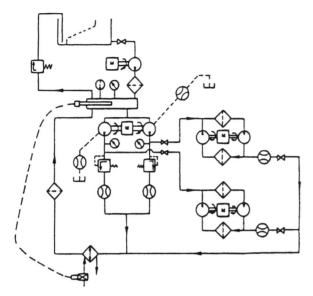

Figure 19 - Gear pump test circuit used at the British National Coal Board.

integrity, monitoring the torque throughout the test would provide greater insight into overall pump integrity during the test.

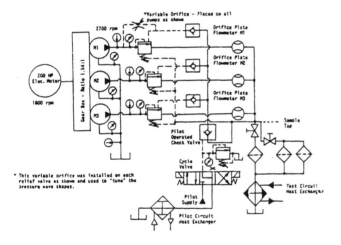

Figure 20 - Wanke gear pump test stand.

4. Evaluation of Pump Lubrication Results

The most commonly used analytical methods to monitor hydraulic pump operation have been reviewed in detail.[34] These methods include: flow analysis, volumetric and overall efficiency, cavitation pressure, minimum inlet pressure and others. Johnson has provided a detailed review of pump testing methods to monitor pump durability and hydraulic operation.[37]

Hunt has viewed pump lubrication analysis from the viewpoint of failure correlation.[35] For example, hydraulic pumps are periodically inspected throughout the test for:

- Cavitation of port plate,
- Bearing wear and break-up,
- Piston slipper pad wear and blockage,
- Blockage of control devices,
- Piston and cylinder wear,
- Case seal defects,
- Gear teeth wear or fracture.

In addition to inspection, temperatures, pressures and flow throughout the test should be monitored since they are indicative of friction generation, fluid contamination. Also, wear analysis and vibration analysis as continuous monitors of wear should be performed.[35,36,38] Common vibrational and acoustical monitoring procedures and interpretive methods have been reviewed previously.[39,40]

CONCLUSIONS

An overview of the various vane, piston and gear pump testing circuits and procedures from the viewpoint of *tribological testing* have been provided here. With the exception of the various tests based on the Vickers V-104 vane pump and perhaps the Hagglunds-Dennison HF-0 piston pump test, none of the pump tests described here have achieved broad industry acceptance. Most of the tests are pass/fail tests based either on inspection for cavitation and wear, fluid flow leakage, or weight loss of critical rotating components. Most of these tests do not employ continuous monitoring of hydraulic efficiencies and pressures, torque, vibrational analysis, etc. Therefore, there is a need to update most of the procedures reviewed above to incorporate these monitoring methods in order to better model the tribological performance of the pump during use with a particular hydraulic fluid.

REFERENCES

1. G. Silva, <u>SAE Trans.</u>, **99**, (1990), p. 635-652.

2. R. Blanchard and L.R. Hulls, <u>RCA Engineer</u>, **20 (6)**, (1975), p. 36-39.

3. A. Hibi, T. Ichikawa and M. Yamamura, <u>Bull. of JSME</u>, **19**, (1976), p.179-186.

4. H. Ueno and K. Tanaka, <u>Junkatsu (J. Japn. Soc. Lub. Engrs.)</u>, **33**, (1988), p.425-430.

5. W.M. Shrey, <u>Lubr. Eng.</u>, **15**, (1959), p. 64-67.

6. R. Renard and A. Dalibert, <u>J. Inst. Petr.</u>, **55**, (1969), p.110-116.

7. G.C. Knight, "The Assessment of the Suitability of Hydrostatic Pumps and Motors for Use With Fire-Resistant Fluids", <u>Rolling Contact Fatigue: Perform. Test. Lubr., Pap. Int. Symp.</u>, Eds. R. Tourret and E.P. Wright, (1977), p.193-215.

8. O.P. Lapotko, V.M. Shkol'nikov, Sh. K. Bogdanov, N.G. Zagorodni, and V.V. Arsenov, Chem. and Tech. of Fuels and Oils, 17, (1981), p.231-234.

9. R.K. Tessmann and I.T. Hong, SAE Technical Paper Series, Paper Number 932438, 1993.

10. J.M. Perez, R.C. Hanson and E.E. Klaus, Lubr. Eng., 46, (1990), p.248-255.

11. K. Mizuhara and Y. Tsuya, Proc. of the JSLE Int. Tribol. Conf., (1985), p. 853-858.

12. L.D. Wedeven, G.E. Totten and R.J. Bishop Jr., SAE Technical Paper Series, Paper No. 941752, 1994.

13. J.R. Hemeon, Appl. Hydraulics, (1955), August, p.43-44.

14. W.H. Reiland, Hydraulics & Pneumatics, March, (1968), p.96-98.

15. J.J. Weaver, Lubr. Eng., 21, (1965), p. 12-15.

16. V.V. Arsenov, L.O. Sedova, O.P. Lapotko, L.V. Zaretskaya, V.I. Kel'bas, and V.M. Ryaboshapka, Vestnik Mashinostroeniya, 68, (1988), p.32-33.

17. "Pump Test Procedure for Evaluation of Antiwear Fluids for Mobil Systems", Vicker's Form No. M-2952-S.

18. This test procedure was communicated verbally by Mr. Paul Schacht as the standard cycled pressure vane pump test recommended by (Robert Bosch) Racine Fluid Power, Racine, WI.

19. J.F. Maxwell, S.E. Schwartz and D.J. Viel, ASLE Preprint, No. 80-AM-713-2, (1980).

20. This is the so-called "Sundstrand Water Stability Test" described in Sundstrand Bulletin 9658. The test Protocol described was conducted by Southwest Research Institute in San Antonio, Texas.

21. G.E. Totten and G.M. Webster, "High Performance Water-Glycol Hydraulic Fluids", Proceed. of the 46th Natl. Conf. on Fluid Power, March 23-24, 1994, p.185-194.

22. S. Lefebvre, "Evaluation of High Performance Water/Glycol Hydraulic Fluid in High Pressure Test Stand and Field Trial", Presented at STLE Annual Conference, May 1, 1993, Montreal, Canada.

23. This test procedure was verbally recommended by Mr. Hans Melief (The Rexroth Corporation, Leheigh Valley, PA). The test procedure is being studied by the ASTM D.02 N.07 committee.

24. "Conduct Test-To-Failure on Hydraulic Pumps", Vickers AA 65560-1SC-4), NTIS No. AD 602244, Sept. 1963.

25. K. Janko, J. Synth. Lubr., 4, (1987), p.99-114.

26. C.M. Edghill and A.M. Rubery, "Hydraulic Pumps and Motors - Development Testing: Its Relationship With Field Failures", First European Fluid Power Conference, Paper No. 31, Sept. 10-12, 1973.

27. V. Hopkins and R.J. Benzing, Ind. Eng. Chem. Prod. Res. Develop, 2, (1963), p.71-78.

28. L.J. Gschwender, C.E. Snyder Jr., and S.K. Sharma, Lunr. Eng., 44, (1987), p.324-329.

29. H.W. Thoenes, K. Bauer and P. Herman, "Testing the Antiwear Characteristics of Hydraulic Fluids: Experience with Test Rigs Using a Vickers Pump", IP Int. Symposium, Performance Testing of Hydraulic Fluids, Oct. 1978, London, England, Ed. R. Tourret and E.P. Wright, Published by Heyden and Son Ltd.

30. C.G. Paton, W.B. Maciejewski, and R.E. Melley, Lubr. Eng., 46, (1990), p.318-326.

31. R.H. Frith and W. Scott, Wear, 172, (1994), p.121-126.

32. G.J. Toogood, "The Testing of Hydraulic Pumps and Motors", Proc. Natl. Conf. Fluid Power, 37th, 35, (1981), p.245-252.

33. T. Wanke, "A Comparative Study of Accelerated Life Tests Methods on Hydraulic Fluid Power Gear Pumps", Proc. Natl. Conf. Fluid Power, 37th,, 35, (1985), p.231-243.

34. American National Standard, "Hydraulic Fluid Power - Positive Displacement Pumps - Method of Testing and Presenting Basic Performance Data", ANSI/B93.27-1973.

35. T.M. Hunt, Technical Diagnostics, Nov. 17-19, (1981), p.89-99.

36. T.M. Bashta and I.M. Babynin, Soviet Engineering Research, 3 (5), (1983), p.3-5.

37. K.L. Johnson, Proc. Natl. Conf. Fluid Power, 30th, 28, (1974), p.331-370.

38. G.A. Avrunin and G.N. Bakakin, Soviet Engineering Research, 9(10), (1989), p.37-39.

39. G.E. Maroney and E. Fitch, 3rd International Fluid Power Symposium, Paper c5, May 9-11, 1973, p.C5-81-C5-96.

40. M. Dowdican, G. Silva and R.L. Lowery, Oklahoma State Univ. - Fluid Power Research Center, Report No. OSU-FPRC-A5/84, 1984.

Positive Displacement Calibration for Laboratory Flowmeters

Peter E. Lucier, Frank J. Fronczak, and Norman H. Beachley
University of Wisconsin - Madison

ABSTRACT

Positive displacement flowmeters can be used to simply and accurately calibrate common flow transducers such as axial turbine and target flowmeters. Two means of utilizing positive displacement devices were studied for use as a laboratory flowmeter calibration. The first method employed a fixed displacement axial piston motor. This proved unsatisfactory due to the difficulty in quantifying flow losses. The second method used a large hydraulic cylinder. An optical encoder measured the position of the cylinder rod, permitting a direct calculation of the flow through the in-line flowmeter being calibrated. Because cylinder leakage is virtually zero at low pressure, flow can be readily calculated knowing the effective cylinder diameter and piston velocity. The method described in this paper permits flow rates to be measured with an accuracy of ± 0.1% of the volumetric flow rate. This paper discusses details of the design of the flowmeter and calibration method.

INTRODUCTION

Many methods have been developed to measure fluid flow in a conduit. Devices used to measure the flow can be divided into two classes: volumetric flowmeters (volume/time, e.g. m³/s) and mass flowmeters (mass/time, e.g. kg/s). Volumetric flowmeters are traditionally used in the testing and analysis of fluid power systems since the components, usually hydraulic pumps and motors, are positive displacement devices. Since pumps and motors move volumes of fluid, it makes sense to measure volumetric flowrates.

While volumetric flowrate and mass flowrate are related, they are not directly proportional to one another. The density of the fluid, and thus the relationship between volumetric and mass flow rates, is affected by system pressure and temperature, both of which can vary widely throughout a fluid power system. For a given mass flowrate, the volumetric flow into a pump will be greater than the volumetric flow out of the pump due to the compressibility of the fluid. When comparing flowmeters, it is important to decide if mass flow principles or volumetric flow principles will be used.

Axial turbine flowmeters are commonly used in the testing and analysis of fluid power systems. They provide the advantages of quick response times, relatively large measuring range with high resolution, and low cost. The output of axial turbine flowmeters is nonlinear over a large flow range as a result of turbine flowmeter design. In theory, the blades of an axial turbine are displaced as a unit volume of fluid passes through the conduit housing the rotor thereby causing the turbine to rotate. Knowledge of the conduit diameter and the speed of the turbine can be translated to a volumetric flowrate:

$$Q = Kn$$

where Q is the volumetric flowrate (cm³/sec), K is the flowmeter constant (cm³/pulse), and n is the turbine rotor angular velocity output (pulse/sec).

Under turbulent conditions, the velocity is uniform across the area of the rotor. In this case, flowrate is linearly related to rotor speed since the relationship is dominated by kinematic effects. When the flowrate is non-turbulent (i.e. laminar or in transition), viscous effects of the fluid dominate and the velocity across the rotor area is non-uniform. The relationship between volumetric flow and turbine speed then becomes non-linear. Figure 1 shows the relationship between K and flow region.

The range of a turbine flowmeter is often maximized by using it in the laminar, transition, and turbulent flow regions.

As can be seen in Fig. 1, three curve fits are required to match flowmeter output to the volumetric flow.

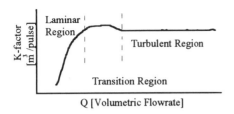

Figure 1: Representative Graph of the Axial Turbine Flowmeter Output for Different Flowrates.

To insure flow meter consistency, it was determined that a primary calibration standard was necessary for our laboratory experiments. Unfortunately, flowmeters generally must be returned to the manufacturer for primary calibration. It is desirable to calibrate the flowmeters before each test setup in order to ensure accuracy. Secondary calibration against other flowmeters has not provided the necessary calibration resolution. Calibration is critical since flowmeters can provide the largest measurement error in the analysis of fluid power systems (1). A calibration standard would both increase the reliability of laboratory experiments and minimize down time for meter recalibration.

It was decided that the calibration method must accurately calibrate flowmeters to 0.1% of the measured flow value, accurately measure flowrates in all flow regions from 60 cm³/sec to 3800 cm³/sec (1 GPM to 60 GPM), and allow for at least 3 seconds of sampling time for a computer to read the flowmeter output. The meter must also be relatively insensitive to fluid viscosity, to disturbances in the inlet and outlet flow profiles, and allow for calibration at operating pressures. A volumetric flowmeter capable of acting as a primary calibration standard for turbine flowmeters was developed, using conventional hardware, at UW-Madison.

FLOW MEASUREMENT TECHNIQUES

Flowrate calibration depends on either a standard of volume and time or mass and time. Primary calibration is generally based on the establishment of steady flow through the flowmeter to be calibrated and subsequent measurement of the volume or mass flow over an accurately timed interval (2). Secondary calibration could involve the use of a calibrated flowmeter to calibrate another flowmeter with lower accuracy demands. Flow measurement methods based on hydrodynamic principles, such as venturi and Coriolis flowmeters, will not be discussed in this paper.

The mass flow measurement method shown in Fig. 2 is based on dynamic weighing and is similar to the method used by the National Institute of Standards and Technology to calibrate flowmeters. This technique involves pumping fluid through the flowmeter to be calibrated and into a reservoir mounted on a load cell. With the system operating at a steady flowrate, the reservoir is continuously weighed. The rate of

weight change with respect to time can be converted into a mass flowrate (or a volume flowrate with knowledge of the fluid density). This method is generally accurate to 0.1% of full scale (2).

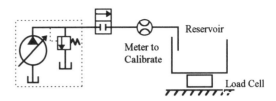

Figure 2: Mass Flow Calibration Method Employing Dynamic Weighing.

A second primary calibration method employs a positive displacement flowmeter to directly measure the volume of the fluid after it flows through the flowmeter to be calibrated. Commercial positive displacement flowmeters can be expensive and delicate for laboratory use. Figures 3 and 4 show two alternative positive displacement flowmeters: a positive displacement hydraulic motor and a hydraulic cylinder. The accuracy of such meters depends on precise quantification of both the internal and external leakage losses. A single hydraulic cylinder is the most efficient of positive displacement hydraulic actuators in terms of friction and flow losses.

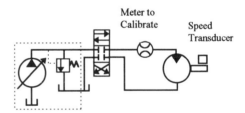

Figure 3: Fixed Displacement Motor as Positive Displacement Flowmeter.

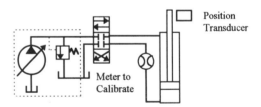

Figure 4: Hydraulic Cylinder as Positive Displacement Flowmeter.

A third calibration method that is commonly used in laboratories involves mounting two flowmeters in series. The first dedicated meter is used to calibrate the second meter, as shown in Fig. 5. Unfortunately, this method does not eliminate the need to calibrate the dedicated meter, and a number of dedicated meters are necessary to calibrate meters over a large flow range due to flowmeter nonlinearities. The accuracy of the meter to be calibrated will always be less than that of the meter used to calibrate.

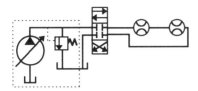

Figure 5: Secondary Flow Calibration Using A Dedicated Flowmeter.

Development of a flowmeter to be used as a primary calibration standard at UW-Madison explored the use of both of the positive displacement methods mentioned above.

HYDRAULIC MOTOR AS A FLOWMETER

Initially, a fixed displacement, bent axis, axial piston hydraulic motor was utilized as a positive displacement flowmeter. The motor's displacement and shaft speed were used to determine the volumetric flowrate. The system pressure was relatively low, near 350 kpa (50 psi.), so the pressure drop and corresponding cross-port leakage across the pump were assumed to be negligible. In the case of a bent axis piston type motor, the overall volumetric are the sum of fluid compressibility losses, cross port leakage, and leakage to the case. The necessary data to calculate these losses to the degree of precision necessary for this project was unavailable and difficult to determine experimentally. This led to the utilization of a hydraulic cylinder as a primary calibration standard.

HYDRAULIC CYLINDER AS A FLOWMETER

Hydraulic cylinders are made to close tolerances that are easily verified, and have almost zero leakage. Flow is determined by measuring the displacement of the piston over time. This can be accurately accomplished with the use of an LVDT (Linear Variable Differential Transformer), linear potentiometer, or an optical encoder. The hydraulic cylinder must be large enough to accommodate the total volume of flow accumulated during each calibration. This includes the volume accumulated during the data acquisition time period and the time necessary for the fluid and cylinder to reach steady-state operation after a quick valve opening.

The single cylinder meter circuit that was investigated is shown in Fig. 6. A cylinder with a 178 mm (7 in.) bore and a 710 mm (28 in.) stroke was selected. This meter is capable of handling flows up to 3530 cm^3/sec (56 GPM) while maintaining a 5 second run time. The 5 second run time includes a 1 second startup period followed by a 3 second data collection period and a 1 second stopping period. This method isolates the sampled data from transients resulting from valve opening and closure and the inertial effects of the piston and rod. The cylinder flowmeter is best suited for steady-state flows. The relatively large inertia of the piston and rod makes this device less suitable for transient flow measurements.

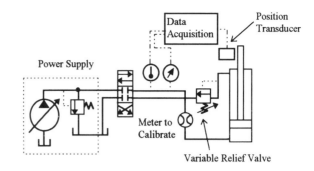

Figure 6: Schematic Layout of Flowmeter Calibration

An optical encoder is used to provide measurement of the piston position. The optical encoder was chosen to facilitate data acquisition and minimize cost. An optical encoder provides an economical alternative to long stroke LVDTs and linear potentiometers. The encoder used was a 2500 counts per revolution single output encoder that is accurate to ± 1/2 count.

As shown in Fig. 7, the encoder shaft is attached to a high precision bushing with a 38 mm (1.5 in) diameter resulting in an effective resolution of 0.05 mm/count (0.002 in/count). A pair of pulleys is used to maintain no-slip contact between the bushing and the Kevlar line attached to the end of the cylinder rod. A 6.7 N (1.5 lbf) free weight maintains constant tension, without noticeable stretch, in the line.

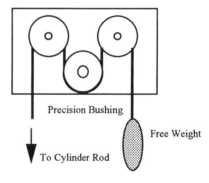

Figure 7: Detail of Encoder and Pulley Arrangement.

A three-position, four-way solenoid valve is used to control the flow into and out of the hydraulic cylinder. The valve operates with relatively slow opening and closing times, on the order of 1 second, to minimize any pressure transients from rapid valve opening and closure. Computer control of the solenoid valve is provided by the data acquisition interface. Since the solenoid valve is not used to meter the flow into the flowmeter, flowrate control is performed at the power supply by varying the displacement of the power supply output pump. The complete calibration setup is shown in Fig. 8. Note that the meter to be calibrated is plumbed after the valve to eliminate any losses from cross port leakage. A regulating valve (variable relief valve) is located on the end of the circuit to control the operating pressure of the system to provide flowmeter calibration at pressures that reflect operating conditions.

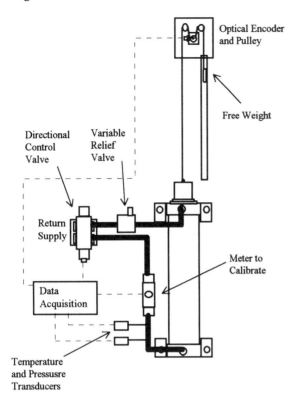

Figure 8: Detail Drawing of Cylinder Flowmeter

RESOLUTION/ACCURACY

As designed, the resolution of the optical encoder, pulley, and Kevlar line is 0.127 mm (0.005 in.). Over the 61 cm (24 in.) stroke of the cylinder used to measure flow, this resolution can result in an error as much as 0.021%. Using dimensional data provided by the cylinder manufacturer, tolerance of the hydraulic cylinder tube is +0.025/-0.075 mm (+0.001/-0.003 in.) which can result in an error in volume measurement as much as 0.057%. Combining these two errors, the maximum overall error in volume reading can be as much as 12 cm^3 (+/- 0.73 in^3). The resulting flow measurement error is 0.078% or less.

For the purposes of establishing a primary calibration standard that was accurate to 0.1% of the measured volume value, additional resolution was not necessary. The error due to bore uncertainty could have been further reduced if the hydraulic cylinder were disassembled and bore measurements were taken. The accuracy of the displacement measurement could be improved if a higher count encoder and/or a smaller bushing on the encoder shaft were used. Both of these actions would increase the number of counts per unit length. Also, to account for cylinder bore variations under pressure, the volumetric error could be determined analytically by the use of thick wall cylinder equations.

Piston seals virtually eliminate any leakage across the piston. Leakage is also minimized since the pressure differential across the piston is very small, just enough to overcome friction. The pressure differential is independent of operating pressure, so the leakage rates are the same for all operating pressures.

The displacement measurement of the cylinder flowmeter was calibrated to an external standard to account for any implementation errors. This was necessary since the Kevlar line runs over a precision pulley. As the line wraps around the pulley, the inside compresses while the top of the line is put in tension. As the line rolls off the pulley, it straightens out. While there was not any appreciable slip between the line and pulley, it is impractical to accurately determine the effective diameter of the pulley and line. Using a large Vernier caliper, the displacement measurement of the pulley and encoder was found to be within 0.2 mm (0.008 in.) of the measured distance with an average difference of 0.1 mm (0.005 in.) These errors were well within the range desired for our laboratory calibrations.

CONCLUSIONS

The need for volumetric flowmeter calibration is adequately met by the use of a double acting hydraulic cylinder. Proper selection of cylinder size and position sensor resolution can provide a very accurate primary steady-state laboratory calibration standard. Using a single cylinder stroke has proven sufficient for meter calibration. Piston reciprocation can result in unwanted dynamics in the flow measurement. Transients from quick valve opening and closing should be allowed to dissipate before flow measurements are taken. This cylinder flowmeter provides an accurate and relatively inexpensive means of routine primary calibration of laboratory fluid volumetric flowmeters.

REFERENCES

1.   Babbitt, G., Lumkes, J., Lucier, P., Fronczak, F., Beachley, N., "Regenerative Testing of   Hydraulic Pump/Motor Systems," SAE Paper 941750, 1994.
2.   Doebelin, E. O., Measurement Systems, Application and Design, 3rd. ed., McGraw-Hill Book   Company, New York, New York, 1983.

A New Technique for Improved Performance of the Pulse Width Modulation Control of Hydraulic Systems

Y. Hou
Caterpillar Inc.

G. J. Schoenau and R. T. Burton
University of Saskatchewan

ABSTRACT

Pulse width modulation (PWM) has been used to alter the performance of on-off hydraulic control valves to make them perform as proportional type flow control valves. Nonlinear performance resulting from time delays in valve switching as well as valve wear due to continuous cycling continue to persist as operational problems. This paper examines a new technique called modified PWM control. The method was found to provide accurate control with a minimum of valve chatter.

INTRODUCTION

In digital control systems, continuous-time signals applied at the input are converted into discrete-time signals by a sampling operation using a sampler. In general, the sampler converts a continuous-time signal into a discrete-time signal according to one of many possible sampling schemes. Pulse width modulation (PWM) is one such sampling scheme. It is a technique that has been successfully used to convert simple on-off flow control valves into analog valves of servo valve flow control accuracy.

The input/output characteristics showing the resulting PWM waveform response to a continuous or analog input is shown in Fig. 1. The output is a high frequency square wave pulse train. The direction or sign of the pulse train corresponds to the direction of the analog input. The width of the pulses is proportional to the magnitude of the input signal. The frequency of the pulse train must be high enough to be effectively filtered by the system connected to the valves.

A schematic diagram of the PWM electrohydraulic control system employed in this study is shown in Fig. 2. Basically, the system consists of a microcomputer, four high speed on-off valves (V1,V2,V3 &V4) and a hydraulic cylinder with feedback sensors. The analog input and feedback signal are converted into a digital signal by a D/A converter and input to the microcomputer that is used as a controller to accomplish the calculation and modulation of the signal. The controller outputs two PWM signals, u_1 and u_2, which are determined by a certain control and modulation strategy via control algorithm software. These two modulated control signals are fed into a high speed on-off valve amplifier which drives four high-speed on-off valves. Each valve is actuated "on" or "off" to control the inlet and outlet flow to both sides of the the cylinder.

The signal, u_1, operates valves V1 and V4 which control the inlet flow to the left side of the chamber and the outlet flow from the right side of the chamber, simultaneously. In a likewise fashion, the signal, u_2, operates valves V2 and V3 which control the inlet flow to the right side of the chamber and the outlet flow from the left side of the chamber. In the ideal case, the flow through the valves would have wave forms similar to as u_1 or u_2. The amount of flow through each valve is dictated by the valve maximum flow rate and the total time the valve is on, t_a, per period, T. That is, over i-th period, the amount of fluid delivered through the valve in one pulse width is $q_v \cdot t_{ai}$. The average flow rate in the i-th period would be $Q = q_v \cdot t_{ai}/T$. By varying the pulse width, the average flow rate can be varied. The output position transducer is used to feed back the actuator position for closed loop control.

PREVIOUS WORK

Central to the operating range of a PWM control system is the speed of response of the on-off valves. Ideally, the valves should open (full flow) and close (zero flow) at a rate at least as fast as the input signal demands. Usually, this is not possible. The development of new valve configurations and associated power amplifiers has resulted in fast switching valves [1,2]. However, a time delay in

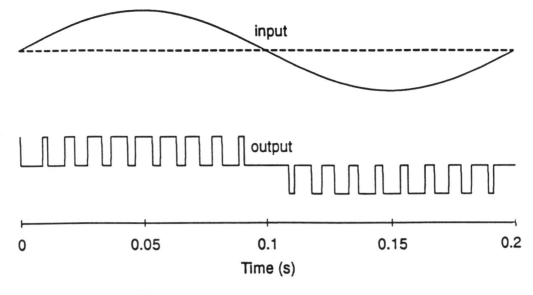

Figure 1 The Operating Principle of PWM

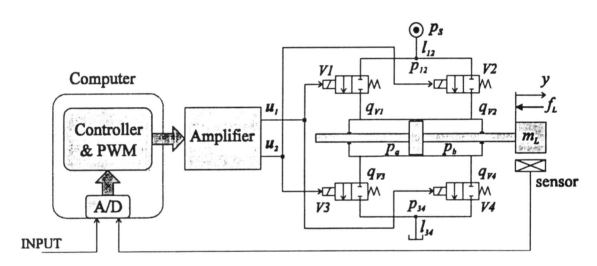

Figure 2 A Hydraulic PWM Position Control System

opening and closing the valve is still present. Although it cannot be completely eliminated, some measures have been taken to reduce the effect.

One approach by Tanaka [3] is based on the introduction of a lead time factor which partially compensates for the influence of the valve delay. Compensation is achieved by adding an offset or lead time to the error signal, thereby increasing the "on" time of the control signal. However, the decay delays in turning the valve off still remain, introducing a discontinuous gain or nonlinearity near the system equilibrium point. A differential PWM control scheme was proposed by Muto et al [4] which eliminated this region of nonlinearity. Their technique is based on having modulated signals simultaneously present at both inputs, u_1 and u_2. The

direction which the actuator in Fig. 2 moves depends on the relative pulse width magnitudes of the two pulse trains. Since these pulses are in opposite directions, only the difference in pulse widths is seen by the system. This effectively enables the generation of any pulse width, greatly minimizing the effect of valve response delays. However, the operation is based on continuous valve cycling, even under conditions of zero error or input signal, significantly reducing valve life. System hydraulic "stiffness" is also reduced with operation in this configuration.

A MODIFIED PWM CONTROL METHOD

A new technique, termed the modified PWM control method, is proposed which increases control accuracy and system stiffness. This method is similar in principle to the lead PWM method for large input signals such as a step input (ex: u_1 receives a pulse train, u_2 is inactive). System response to a step input is shown in Fig. 3.

When the error signal becomes small, as the piston approaches the objective, the amplitude of the error signal is monitored in the error signal monitoring algorithm to determine if the error is small enough to turn on the compensator. Normally, in the small error range, the pulse width of the pressure difference, Δp, cannot follow the control pulse train, u_1, in a manner (proportional) to the amplitude of error. This is because t_{on} and t_{off} (the rise and decay delays of the valves) take a large portion of the pulse width of Δp. With the compensator on, the modulator generates a compensation pulse train, u_2, with a fixed width of t_{on}^+. At the same time, the width of the control pulse train, u_1, will be increased by the period of t_{on}. u_2 creates a back pressure on the piston which just balances the t_{on} and t_{off} portion of the pulse width of the pressure difference, Δp, generated by u_1. Thus the pressure difference, Δp, created by the signals, u_1 and u_2, will be applied in a proportional manner to the amplitude of error over the whole duty cycle. The effectiveness of the modified PWM also can be predicted in the response of the normalized mean pressure difference (shown in Fig. 4) which demonstrates a relative linear characteristic near the equilibrium point.

Once the system reaches the desired position (the error falling within a desired range), the monitor locks the system by turning off the PWM which significantly reduces valve noise and increases the life span of the on-off valves.

SIMULATION RESULTS

Simulations of all three system types were conducted to verify the superiority of the modified PWM control approach. The systems simulated included, in addition to the proposed modified PWM method; the lead compensation PWM method of Tanaka [3] and the differential PWM control scheme proposed by Muto et al [4].

The simulation study was based on the simple proportional feedback system of Fig. 2. The load consisted of a mass, m_2, of 10 kg and a force, f_2, of 100 N. The PWM frequency was 500 Hz and the system natural frequency was 70 Hz. The relatively high PWM frequency was chosen based on a study using a prezoelectric actuator valve [5]. The system was being operated at a supply pressure, P_s, of 10 MPa.

All systems were subjected to an identical step input in the desired displacement, U. Figure 5 shows the resulting load output displacement, Y, for each system. The corresponding pressures on either side of the piston, P_a, and P_b, are shown in Fig. 6.

As shown in Fig. 5, the speed of response (rise time) is basically the same for each system. The main difference is in the steady state characteristics. The lead compensated method shows a high degree of oscillation, which is also very apparent in the pressure signals (Fig. 6). This is a consequence of the inherent discontinuous gain. The differential PWM technique does not exhibit any oscillation in

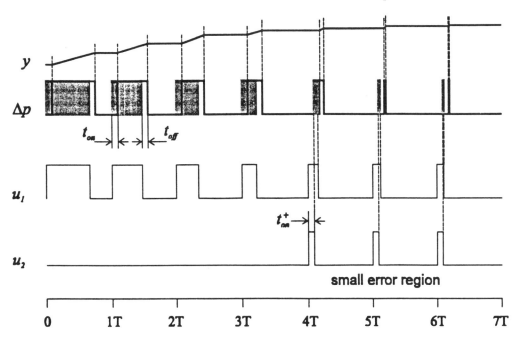

Figure 3 The Operating Principle of the Modified PWM Method

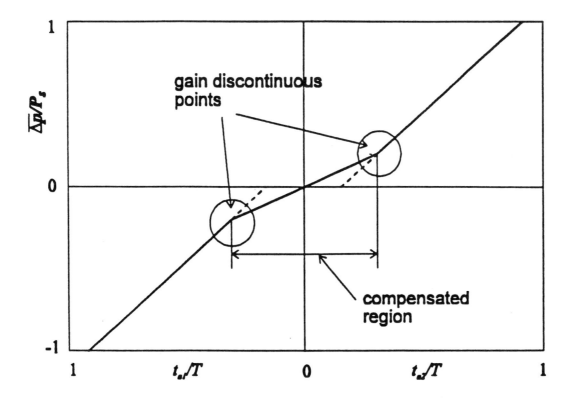

Figure 4 Duty Cycle Gain Characteristic for the Mean Pressure Difference

load output. However, there is continuous valve chatter (pressure oscillations) which would result in excessive valve wear. Also, there is a small steady-state output error because of the low system stiffness. The modified PWM method achieves the most accurate steady-state displacement while eliminating valve chatter under steady-state conditions.

CONCLUSION

A modified PWM control method has been devised which offers advantages over existing techniques. A more accurate steady-state response is possible with a reduced amount of valve wear due to less cycling of the on-off valves.

ACKNOWLEDGEMENTS

The authors are grateful for the technical assistance of Mr. D. Bitner of the Department of Mechanical Engineering Fluid Power Laboratory, University of Saskatchewan and for the financial assistance in the form of grants from the Natural Sciences and Engineering Research Council of Canada.

REFERENCES

1) Y. Hou, "Development of a Digital Electrohydraulic Control System and a High Speed On-off Solenoid Valve", M.Sc. Thesis, Beijing Institute of Technology, Beijing, China, 1988.

2) Y. Hou and F. Cao, "A Digital Position System Using High Speed On-Off Solenoid Valves", The 2nd National Fluid Power Symposium, Shangsha, China, 1989.

3) H. Tanaka, "Electrohydraulic Digital Control of 3-Way On-Off Solenoid Valves", Transactions of Japanese Society of Mechanical Engineers, Vol. 50, No. 458, pp. 2663, 1984.

4) T. Muto, H. Yamada and Y. Suematsu,"Digital Control of a Hydraulic Actuator System Operated by Differential Pulse Width Modulation," JSME International Journal, Series III, Vol. 33, No. 4, pp. 641, 1990.

5) Y. Hou, "A Pulse Width Modulation Based Control System using High Speed On-Off Valves," M.Sc. Thesis, Dept. of Mechanical Engineering, University of Saskatcheawn, 1994.

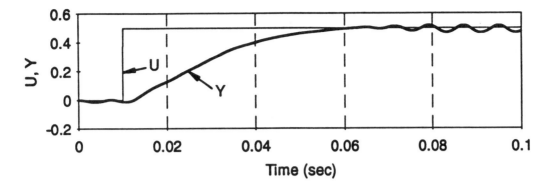

a) Lead Compensted PWM Method

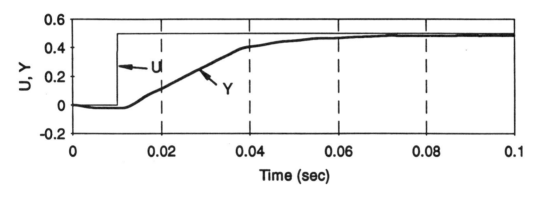

b) Differential PWM Method

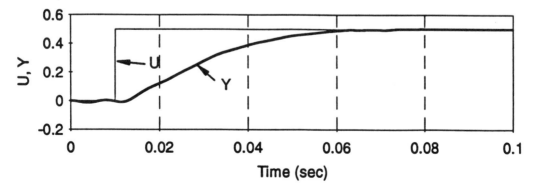

c) Modified PWM Method

Figure 5 Displacement Step Response Characteristics for the Various PWM Methods

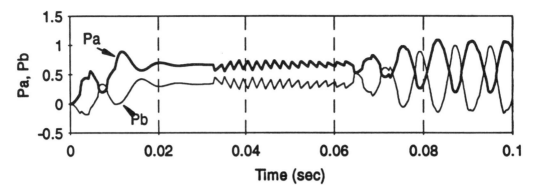

a) Lead Compensted PWM Method

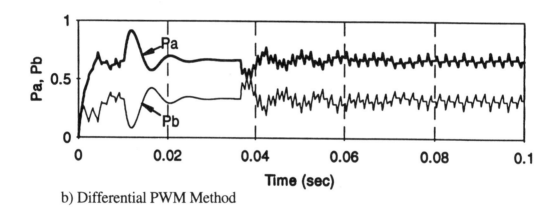

b) Differential PWM Method

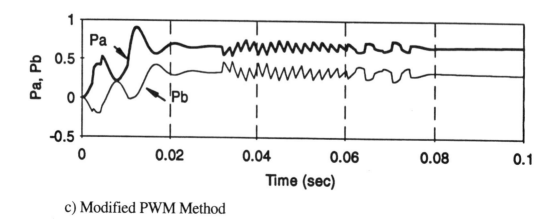

c) Modified PWM Method

Figure 6 Cylinder Pressure Step Response Characteristics for the Various PWM Methods

Development of a New Variable Displacement Axial Piston Pump for Medium Power Applications

Jeffrey A. Baldus and David Wohlsdorf
Sauer Sundstrand

ABSTRACT

This paper describes the development, function, and design features of the new Series 45 open circuit hydraulic pump. The design offers increased performance with significantly reduced size, fewer components, and increased modularity compared to competitive products.

DEVELOPMENT

In 1992 Sauer-Sundstrand decided to design and develop a family of variable displacement open circuit pumps targeted specifically at mobile applications. Based on past successes of other product designs, a multidisciplinary team was formed to apply the tools of Quality Function Deployment (QFD) and Design for Manufacturing (DFM) to develop a unique product which met the needs of both external and internal customers.

Members of the team visited and surveyed North American and European manufacturers of mobile equipment to determine their current and future open circuit pump needs. QFD provided a structured process of weighting and bench marking those needs for maximum benefit and then correlating them to specific product characteristics. Several needs were expressed universally :

- Low Noise Levels
- Compact Size
- Simple without compromising function
- Leak Free
- Low Installed Cost
- Fast and Stable controls

In addition to meeting the needs of the external customer, it was recognized that Series 45 must also meet the needs of the internal customers - namely manufacturing, assembly, and service.

Representatives from these areas formed a cross functional team that was given the charter of incorporating Design for Manufacturability in the Series 45 program.

The team identified several fundamental design principles that would have the greatest impact on the overall manufacturability, including:

- Minimizing the number and cost of "Model Maker" components for increased flexibility
- Minimizing the number of threaded fasteners and press fits to aid assembly and service
- Minimizing the total number of components to reduce cost and improve reliability
- Using symmetric parts and designed in poke-yokes to eliminate assembly errors

The cross functional team was also responsible for creating a design baseline from which Series 45 concepts could be benchmarked. A composite "Best in Class" design was established by disassembling and evaluating approximately 20 different pump designs. Detailed comparisons were made of each design's individual parts and features. Those rated the highest were selected as a baseline. Critical characteristics not satisfied by any of the designs, such as size, had more stringent benchmarks established based on the customer requirements expressed during the QFD.

Having actual pump components available at all discussions had the additional benefit of helping facilitate in-depth manufacturing and assembly questions before a concept or any engineering prints had been created. More importantly, it promoted simultaneous development of the manufacturing process and design while avoiding the pitfall of duplicating "artifacts" from previous designs.

The output of the QFD and DFM activity was a product specification driven by both external and internal customers. From this specification concepts were developed and benchmarked against the theoretical "Best in Class" design. Several iterations were required to create a concept capable of meeting the requirements.

The concept that was ultimately pursued was extremely compact and contained an integrated control. This made a thorough engineering analysis extremely important due to the major impact any changes would have on the overall concept and the program schedule.

An in-depth analysis of the concept was performed prior to generating detail drawings or hardware. Structural components were analyzed using Finite Element for stress and deflection. The endcap, valve plate, and cylinder block were optimized for maximum filling efficiency and higher speeds.

The control system was studied in detail with dynamic models developed which included the control, pumping elements, and typical load elements. Models using computer generated numerical solutions were created for analyzing the non-linear systems.

The models were useful for verifying the control logic, as well as, optimizing many of the control parameters prior to any testing. As a result, very few control changes were made after the pump was designed.

From these analyses a concept layout was created and suppliers were selected based on the strategic partnering relationships which already existed. The responsible process engineers from all internal and external suppliers were invited in-house for a multiple day design launch. During this meeting, flow charts were created detailing each parts progression from the first step until final assembly into the pump. Having both process and design engineers together and focused helped create an environment that allowed for rapid exchange and iteration of ideas. The notes and sketches from the meeting were used to further refine the concept and reduce the cycle time required to detail the component parts .

A full scale 3-D Laminate Object Model (LOM) of the design was created after the housing and endcap geometry had been defined. This model, along with the performance predictions, were taken to customers who had been originally visited for verification that the design met their needs. Detailing of the individual pump components began only after the completion of this QFD verification.

A sectional view of the Series 45 pump design which resulted is shown in Figure 1. Pump ratings are summarized in Table 1.

Figure 1. Sectional View of the Series 45 Pump

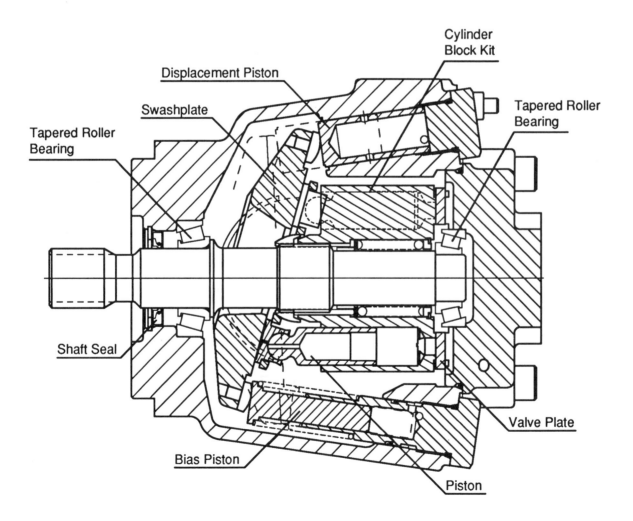

Table 1. Series 45 Technical Specifications

	Dimension	Frame Size 057	
Displacement		cm³	57
		in³	3.48
Input speed	Minimum	min⁻¹ (rpm)	500
	Rated *	min⁻¹ (rpm)	2450
	Maximum *	min⁻¹ (rpm)	3200**
Maximum working pressure		bar	350
		psi	5075
Continuous working pressure		bar	280
		psi	4060
Flow at rated speed		l/m	140.3
		gpm	36.9
Theoretical input torque at maximum flow and pressure		Nm/bar	0.907
		lbf•in/1000 psi	554
Mass moment of inertia of the int. rotating parts		kg•m²	0.0046
		lb•ft²	0.1092
Weight	Axial ports	kg	23.6
		lb	52
	Radial ports	kg	30 (estimated)
		lb	66 (estimated)

* Refer to General Technical Specifications in Series 45 Axial Piston Open Circuit Pumps Technical Information ma

** With pressurized inlet

PUMP OPERATION

The pump is an axial piston design with a cradle type swashplate resting on polymer cradle bearings. Two control pistons act on the swashplate. The bias piston uses pump outlet pressure and spring force to bias the swashplate to full displacement (maximum angle). The servo piston, which is larger in area, responds to the control to overcome the bias piston force and reduce the swashplate angle. As the shaft turns the cylinder block, the pistons reciprocate in their bores due to the reaction caused by rotating while being held on the angled swashplate. The flow created is communicated to the pump inlet and outlet through the valve plate. An integral control in the pump housing regulates the pump displacement by pressurizing the servo piston to control the swashplate at any angle between its maximum and zero. A variety of controls are available to regulate pump displacement in response to pressure, flow, or torque.

The Load Sensing (Flow and Pressure Compensation) control is shown in figures 2 and 3. When used with a closed center load sensing directional control valve a highly efficient, and controllable system with low noise levels can be achieved. The load sensing (LS) control acts to maintain a constant pressure drop across an external directional control valve. Load pressure (pressure from downstream of the directional control valve) is sensed through the 'X ' port of the pump and acts, in combination with the control spring, on one end of the pump control spool. The other end of the control spool has pump outlet pressure (pressure from upstream of the directional valve) acting on it. Thus the spool senses the pressure drop across the directional valve and will be in equilibrium only when the pressure drop across the control spool (and therefore the directional valve) is equal to the spring setting.

For any given opening of the directional valve, a reduction in the pressure drop across the directional control valve will disturb the pump spool equilibrium. The spring force will move the pump control spool to tank the servo piston and increase the pump until the pressure drop increases to that required for equilibrium. Conversely, if an increase in the pressure drop is sensed, the spool moves against the spring, porting oil into the servo piston reducing the pump flow until the pressure drop is reduced.

Figure 2. Load Sense Control Schematic.

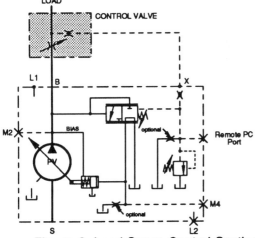

Figure 3. Load Sense Control Sectional View

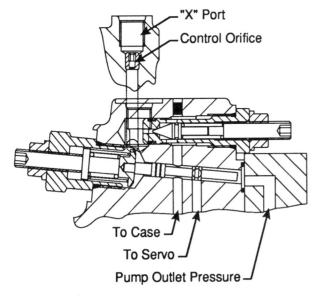

If at any time the load pressure exceeds the pressure compensation setting the PC pilot valve will open creating a delta pressure across the control orifice. This will override the load sensing causing the pump control spool to port oil to the servo piston and de-stroke the pump.

When system flow is not required and the sensing port is tanked, pump pressure will adjust to match the force of the control spring. The pump will standby at zero flow and low pressure (typically 20 bar)

DESIGN FEATURES AND BENEFITS

REDUCED NUMBER OF COMPONENTS - The Series 45 pump has 20-30% fewer parts than comparable open circuit pumps. Key to reducing the number of parts was development of a single piece housing with an integrated control. The housing design eliminates the need for separate side covers, control housings, and the associated screws and seals. The resulting design has no gaskets and uses only 4 screws resulting in a leak free design that is easy to assemble and service.

SIZE - The single piece housing design also significantly reduced the size of the pump. The box volume (length• width • height) is as much as 50% smaller than comparable pumps. In many existing applications with limited space, this makes it possible to replace a smaller displacement pump which has become inadequate due to increased flow demands. Pump width, critical in applications mounted along engines, vehicle frames, or multiple pump drives was minimized by eliminating side covers, bolts and bosses leaving narrow "clean sides" for tight mountings. Outline dimensions are shown in Figure 4.

MODULARITY - Modularity allows tailoring the pump to specific customer needs while minimizing the impact on manufacturing. Optimally, the desire is to reduce the number and cost of so-called "Model Maker parts" without limiting the number of options available to the customer.

The modularity of traditional open circuit pump designs was severely compromised by the need for separate endcaps for right hand and left hand rotations. The endcap can represent 20% of the pump's total cost and be one of the longest lead time components making it very undesirable as a model maker part. The Series 45 design eliminates this liability by using the same endcap for both directions of rotation. A change of rotation requires only a new valve plate, and indexing the endcap 180°. Similarly, traditional designs have utilized separate, bolt-on controls to satisfy options. This lead to an undesirable increase in the number of parts, number of leak paths, and package size. Market analysis done for the Series 45 showed that the majority of mobile applications are served with either a Load Sensing (LS) control or a Pressure Compensating (PC) control. To facilitate modularity, the control was designed such that the Load Sensing and Pressure Compensating controls were very similar. The Series 45 control uses a single spool and a pilot valve for all control functions. The only hardware differences between the PC and LS controls is the control spool. Traditional control designs use a single spool for the PC control and two separate stacked spools for the LS control necessitating different control assemblies which, due to their cost and lead-times, are undesirable as model maker parts. Table 2. Illustrates the modularity of Series 45 compared to a traditional design.

Figure 4. Outline Dimensions of Radial and Axial Ported Series 45 Pump.

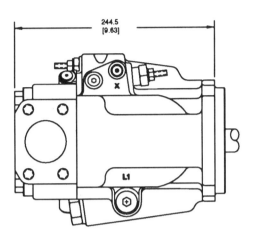 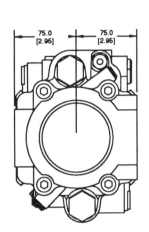 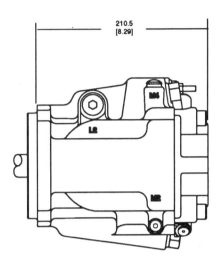

Table 2. Hardware Required to Convert from a CW Pump with an LS Control to a CCW Pump with a PC Control.

	Series 45	Traditional Design
Change of Rotation	CCW Valve plate	CCW Valve plate CCW Endcap New Gasket
Change of Control	Control Spool	Control Assembly Housing Control Spool Spring Spring Guide Spring Seat Adj. .Screw Lock/Seal Nut

NOISE REDUCTION - Both fluid borne and structure borne noise were addressed during the development of Series 45. Reduction of fluid born noise centered on the valve plate. The Series 45 valve plate was first modeled on a computer, and then optimized through frequency analysis of the pump air borne noise and swashplate vibration spectra to obtain a valve plate design that is optimized for noise and efficiency.

Structure borne noise was reduced by the increased stiffness and greater structural integrity of the single piece housing design. Polymer cradle type journal bearings were selected over anti-friction bearings to aid swashplate damping. Noise values measured in dB(A) at 1.0 m (3.28 ft.) from the unit in a semi-anechoic chamber are shown in Table 3. Anechoic levels can be estimated by subtracting 3 dB(A) from these values.

Table 3. Pump Noise Levels

	Sound Level dB(A)			
	210 bar (3000 psi)		280 bar (4060 psi)	
RPM	1800	2450	1800	2450
dBA	76	78	77	80

HIGHER EFFICIENCIES - High efficiency levels were achieved due to the Series 45's compact design and efficient valve plate porting. Deflection of pump components such as the shaft, swashplate and endcap lead to increased internal leakage and reduced efficiency. The compact nature of the Series 45 design inherently reduced these losses by minimizing deflection. This was achieved by reducing the distances between the shaft bearings, the swashplate bearings, and endcap bolts. Overall and volumetric efficiencies are summarized in Figure 5 and Figure 6.

Figure 5. Pump overall and volumetric efficiency.

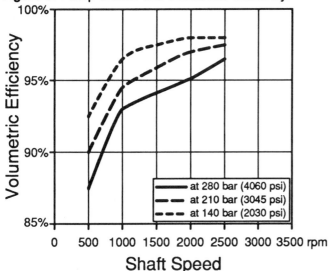

Figure 6. Pump overall and volumetric efficiency.

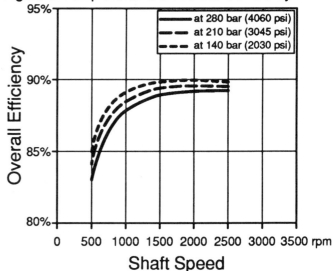

IMPROVED CONTROL PERFORMANCE - Fast and stable control performance was obtained by integrating the control into the housing directly adjacent to servo piston. Control flow has a very short, direct passage to the servo piston and must pass only through 1 spool vs. 2 spools of the traditional design. Series 45 response times (full displacement to zero displacement) are 20-30 mS and recovery times (Zero displacement to full displacement) are 30-40 mS when tested per SAE J745. Additional advantages of this type of control design versus the traditional design of 2 stacked control spools are:

* Small PC spring size (Loads on PC spring reduced by a factor of 10 due to the area difference between the end of a spool and the end of a poppet)
* Reduced contamination sensitivity of poppet valve vs. spool
* Serviceability of pilot cartridge valve design
* Pilot design allows for easy remote actuation and interface with electro-hydraulic controls.
* Ability to optimize system stability by changing gain orifices to tune the system.

IMPROVED SERVICEABILITY - The series 45 design incorporates several features to improve serviceability. The Servo and Bias pistons are accessible without removing the pump endcap allowing them to be serviced without removing the pump from the vehicle.

Service gage ports for system pressure, servo pressure, and case pressure have been added and have identification labels cast into the housing adjacent to the ports. The identification labels match those in the technical literature for ease in diagnostics. Threaded control orifices for fine tuning the control to match load dynamics are located less than 30mm below the surface. Those ports are labeled allowing them to be changed easily without requiring the removal of the control. The control pilot valve is a self-contained cartridge, allowing it to be removed and serviced easily.

LOWER SYSTEM COST - The Series 45 design allows 100% through torque for tandem pump mountings reducing the need for multiple pump drives. Installed costs are further reduced because of the compact size and 4 bolt mounting pad make it unnecessary necessary to support the rear pump of the tandem in most applications.

External plumbing requirements are reduced by eliminating the need for external flow used to flush the case in applications which spend extended periods of time on (high or low pressure) standby. This cooling is accomplished by using the pilot flow from the control and valve plate to cool the pump case.

The fast control response/recovery may allow for size reduction or elimination of accumulators and relief valves in some circuits.

SUMMARY

Utilizing QFD and DFM in the development of the Series 45 pump has resulted in a unique new design which meets the performance requirements for medium power mobile applications while making dramatic improvements in pump size, number of components, and modularity.

Saving Hydraulic Horsepower Using Pneumatic Actuated Clutch in Mobile Hydraulic Systems

R. Douglas MacDonald
Horton Manufacturing Co., Inc.

ABSTRACT

This new clutch design enables machine builders to rest the pump when the circuit does not require any flow.

INTRODUCTION

This paper introduces a simplified approach to save heat, horsepower, and fuel by using a multi-disc clutch to disconnect a hydraulic pump when the circuit is at rest.

This device can be direct driven and mount between a SAE mounting and the pump. This "sandwich" approach allows the designer to close couple the PTO, the clutch, and the hydraulic pump.

TEXT

When the circuit is at rest must the hydraulic pump need to be operating? This question has faced design engineers for years. What they were looking to accomplish was a method to stop the pump when the circuit does not require any flow. When this was not possible a variety of methods of unloading pumps are used. These methods include:

1. By pass circuit - The circuit is blocked and the pump flow either is returned directly back to tank or in some instances the pump output is returned directly back to the inlet of the pump.

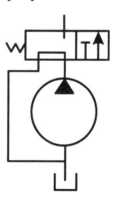

2. Inlet restriction - This method restricts the amount of fluid into pump, allowing enough to lubricate the spinning pump.

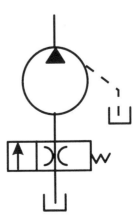

3. Pressure compensated pump - This method requires the use of an expensive variable volume vane or piston pumps.

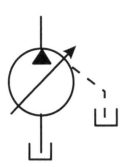

There are many derivatives of the above three types of systems which designers have employed in an effort to conserve power when the circuit is at rest.

All of these require the pump to keep operating even under low load. Designers must still take into account that heat will be generated during these "at rest" times and this generated heat must be dissipated and/or removed.

One method to solve this problem is the use of a pneumatic actuated disconnect clutch.

EXAMPLE

Recently a manufacturer of refuser packer vehicles was faced with the following problem:

Required: 110 liters at 1200 rpm engine speed
Problem: When the vehicle was run at highway speed (approximately 2,400 rpm), the pump would produce approximately 220 liters.
At 220 liters, backpressure built up to 20 bar thru an open center system

$$KW = \frac{bar \times liter}{600} = 7.3 \ KW$$

(Eff. not taken into account)
7.3 KW was generated as heat and must be removed.
7.3 KW requires approximately 2 liters fuel per hour while running
Thus over a year the savings can be substantial.

HOW DOES THIS CLUTCH OPERATE?

The following illustration shows a typical multi-disc clutch

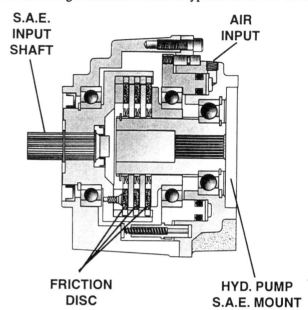

SPECIAL FEATURES OF THE CLUTCH

Self-compensating - as friction plates wear the pneumatic actuator travel increases but the force remains constant.

HYDRAULIC SYSTEM BENEFITS

Noise in hydraulic system - Reduction of the speed of the pump will reduce pump noise. The designer can now size the pump to produce the flow required to operate the machine at optimum. When engine speed increased above a predetermined engine speed the speed is sensed and the clutch is disengaged thus saving fuel and reducing noise.

Component Sizing - This also means the hydraulic components (Relief Valves, Directional Valves, Component Sizing and Hoses) can be sized to match the pump output - 110 liter/min. - not the max flow at highway speed - 220 liters/min. This saves component cost to the vehicle manufacturers.

Pollution Awareness - EPA and state DNR requirements dictate that oil spills shall not occur. Certain manufactures now use this clutch to stop any pumping of oil if a oil leak or hose failure is detected. Other horsepower saving devices or circuits have the pump running constantly therefore, will pump all the capacity of the reservoir on the ground. This requires extensive and expensive oil clean-up.

CONCLUSION

Using a PTO disconnect clutch allows the pump to be at rest while the vehicle is traversing yet be engaged remotely at speed thus saving heat, horsepower, fuel, and extending the life of the pump.

Rapid Prototyping Shortens Hydraulic Component Development Time

Ron Jones, Scott Miller, and Ivan Stretten
Dana Corp.

ABSTRACT

A few years ago hydraulic fluid power component manufacturers had the luxury of long lead times to develop new products. In today's competitive global market, pump and valve design engineers must be able to shorten development lead times and get new, less costly products to production in order to satisfy customer demands. This paper describes how one fluid power component manufacturer uses rapid prototyping technology to speed up the development cycle by making: fit and form models, design evaluation test samples, and tooling for prototype castings.

INTRODUCTION

In the competitive global market of today, manufacturers of hydraulic fluid power components must be able to respond quickly to customer demands for new or improved products. A delay of a few weeks on prototype delivery to an original equipment manufacturer can make the difference between gaining market share and being selected as the production supplier or losing new business to a competitor. Rapid prototyping technology with computers playing a key role, give engineers and designers the ability to quickly respond to customer's needs. Shortened development lead times can be accomplished by generating 3-D surface or solid models for a quick visual analysis. Models for fit or dimensional analysis and functional testing, or tooling for prototype castings and other long lead time components can be quickly produced accelerating the design process.

Many fluid power component manufacturers offer a full line of products that make up a complete system: hydraulic pumps and motors, directional and pressure control valves, and cylinders. Cylinders with their relatively simple geometry and construction of tubes, rods, pistons, glands, and cylinder ends can be manufactured easily from bar stock, forgings, and hydraulic tubing. However, pumps, motors, and valves present a different more difficult problem. Not only do many of these units require special mounting configurations but complex internal passages are required to route oil flow from one cavity to another. In most cases, the only way this can be achieved is through the utilization of ferrous or aluminum castings with the oil passages generated by coring. Unfortunately, the tooling required to make these castings can be complex and require long lead times even for prototype samples. Depending on the design, delivery of sample castings may require several months from the initial design layout. Additionally, pumps and valves use rubber and plastic parts for pressure balance seals, joystick grips and boots, etc. Lead times for these components also require several weeks before sample components can be acquired.

The comparison below illustrates how the lead time of prototype castings can be reduced using rapid prototyping technology. Twelve weeks from layout and design to delivery of samples is not unusual in new product development. The eight week lead-time for prototype casting patterns and core boxes has the most influence on a typical project. By using rapid prototyping technology, prototype tooling can be machined normally within a few days after the completion of the solid models, reducing the project time from twelve weeks to four.

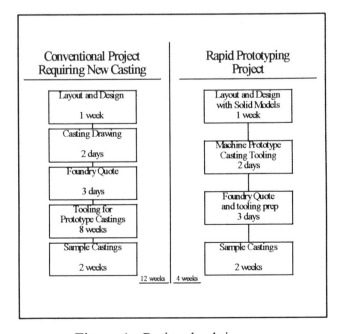

Figure 1: Project lead times

The following is our definition of Rapid Prototyping and a brief description of the technology. Seven rapid prototyping applications are reviewed and then 3 case studies are presented illustrating how development lead time can be shortened on new hydraulic fluid power components.

THE RP PROCESS

Rapid prototyping (RP) is the efficient use of proven CAD/CAM technology to convey engineering concepts to customers, manufacturing engineers, suppliers, and designers. It shortens product development cycles thereby reducing the time-to-market for an advantage in today's competitive market. The technology our rapid prototyping group uses includes a three dimensional computer-aided design (CAD) system linked to a three axis computer numerical controlled (CNC) machining center.

The process begins at an engineering workstation where a 3D CAD model is generated from either an existing engineering drawing, sketches, preliminary drawings or an IGES file from a 2D or 3D database. Parting lines, draft angles and core print information is then added to the CAD model if required, and a computer generated shaded rendering is processed. This realistic rendering is a valuable tool for both the CAD operator and the engineer because it allows everyone involved to visualize the part design before any time is spent machining the prototype. At this time the design is evaluated to determine possible engineering changes or to foresee any difficulties in the manufacturing process.

The next step or the RP process is to generate toolpaths. These toolpaths are the instructions used by a three-axis CNC machining center to guide a cutting tool across a block of high density polyurethane tooling board producing a prototype model or tool to the exact specifications of the CAD model. Toolpaths are calculated by the computer based on parameters defined by the operator. These parameters include cutter size and shape, spindle speed, feed rate, machining direction, and surface finish tolerance. After processing this information, the computer outputs a cutter location (CL) file which is visually verified at the engineering workstation. This allows the operator to see the exact cutting tool path before the program is sent to the machining center.

After verification of the toolpaths, the CL file is ready to be downloaded to the machining center via a personal computer (PC) linked to the engineering workstation through a local area network. Because of the significant size of the CL file (usually 100KB - 1MB in size), it is sent to the machine through a direct numerical control (DNC) connection. This connection is established using a Microsoft Windows (TM) based communications package utilizing the serial port on the PC and a serial port on the machine center controller. With this connection established, the machine tool simply executes the data one line of code at a time without storing any of the CNC code.

Before the toolpath files are downloaded, the raw stock must be secured to the machine tool bedplate. This is accomplished by clamping the stock in a machine vise or by mounting the stock to a piece of machined hardwood using steel dowel pins and threaded fasteners attached from the underside of the material. The hardwood block is then secured to the machine vise. Next, the work offsets must be preset into the memory of the controller. The offsets reference the zero position of the work piece to the origin of the CAD model. The tool length offsets are also set at this time, referencing the position of the tip of each cutting tool relative to the origin of the CAD model. After setting the machine offsets and securing the workpiece to the machine tool, the toolpath files are downloaded to the machining center and the prototype model or tool is machined.

RAPID PROTOTYPING APPLICATIONS

Our rapid prototyping process has been successfully applied to the following applications:

FIT AND FORM PLASTIC MODELS - This application enables engineers to quickly turn preliminary sketches into physical models. It gets the design off the drawing board to where it can be used for clearance studies, assembly checks, or customer evaluation. The engineer can now attend meetings with a part in hand instead of just drawings. If the RP group is involved at the early stages of design, several weeks can be saved by eliminating the need for detailed drawings.

PROTOTYPE TOOLING FOR SAND CASTINGS - After machining a fit and form model, the same CAD data can be used to build prototype patterns and core boxes for producing sample cast metal parts. The first step in any casting project is to work with the foundry and get their involvement early in the tooling design stage. Most foundries will determine parting line location, minimum draft angles, core print and core clearance requirements, and shrink allowances. This information is added to the CAD file and tooling is then machined. If the foundry has automatic molding equipment, the patterns will usually be mounted to a match plate. With hand molding equipment, a loose pattern and parting board is all that is required. With plastic core boxes, the foundry must have a "cold box" core making process. This can limit foundry choices in some cases. Establishing a good working relationship with a prototype foundry is essential in getting sample castings within the desired time frame.

MOLDED URETHANE PROTOTYPES - Plastic and rubber parts can be prototyped using plastic molds. Two part urethanes are gravity fed into molds at room temperature and pressure. Various physical properties including durometer and color can be simulated using this process. Although these parts do not have the same mechanical properties as the production molded pieces, they can prove valuable in the evaluation of a new design. With this process, the tooling cost and lead time can be greatly reduced over the traditional method of making a production injection molding tool and calling the first parts out of the mold a "prototype". If the prototype molded part contains difficult geometry or negative draft areas, silicone rubber molds can be used instead of machined plastic molds. In this case, the silicone is poured over a machined master model to produce the mold cavity.

TOOLING FOR PLASTER CAST PROTOTYPES - Plaster casting is a mean of producing a prototype die cast or permanent mold aluminum component. In this process, the machined master is shipped to a plaster cast foundry where they are used to produce flexible rubber molds. Plaster is mixed and then poured into these rubber tools forming the molds into which the molten aluminum alloy is poured. The rubber molds allow thin sections of plaster to be removed from the mold without breaking. Ribs, slots, thin wall sections, and minimal draft conditions are easily obtained using this process. The plasters are first placed in a furnace and baked overnight to remove any moisture that is present. Because the plaster is destroyed when removing the part from the mold, one set of plasters is required for each aluminum casting. The next day, these plasters are assembled and molten aluminum is poured into the molds. After solidification, the molds are broken apart and the castings are removed, cleaned, and heat-treated if required. Because the plaster mold has a smooth, hard surface, the resulting prototype tends to resemble an actual die cast or permanent mold part. Using this process, lead times as short as one week after the completion of the machined master model can be expected.

CLEAR ACRYLIC ENGINEERING MODELS - Using the RP process, a block of clear acrylic can be machined to the exact specifications of a CAD model. This produces an engineering model which can be used for oil-flow analysis, design verification, or a sales and marketing tool. With only slight adjustments in spindle speeds and feed rates, acrylic models can be machined, drilled, tapped, and bored to exact tolerances. A "flame polishing" technique is used to remove all tool marks to make the part clear again.

METAL "HOG-OUTS" FOR MECHANICAL TESTING - After machining a fit and form plastic model, the same CAD data can be used to produce functional metal prototypes from bar stock. These prototypes can be used for field testing or durability testing in a controlled environment. Again, if the RP group is involved at the early stages of design, several weeks can be saved by eliminating the need for detailed drawings. In many cases, parts have been designed, built, and tested without ever producing a paper drawing.

PRODUCTION TOOLING - This application only works with certain suppliers that have processes compatible with our rapid prototyping methods. Additionally, outside pattern shops and die shops must be involved to complete the production tooling. One example, is the use of a cast iron foundry where the molding process is so closely monitored and controlled that a mounted plastic tool in a fully automated molding line is used. The plastic tooling is actually molded from a RP master model by an experienced pattern shop. These tools last for several thousand parts before the patterns must be reproduced from the original masters. This process is best suited for low volume production runs without complicated core setting. Another example of production tooling benefiting from the RP process is in the forging industry. Hardened die cavities are

formed using an electric discharge machine (EDM). With this process, male graphite electrodes are machined using the RP process. These conductive electrodes are then burned into the die cavity forming the negative impression in the hardened die. The last application of RP models for production tooling involves the machining of a plastic master model for use as a duplicating aid by a production pattern maker. The master model will contain all of the required features including core prints and shrink rules, however, it is only used to duplicate the pattern into a material suitable for production.

CASE STUDIES

The following case studies highlight three of the most common rapid prototyping processes that are used. Each example demonstrates how the RP process is used to dramatically decrease lead times and costs associated with getting a new design to a customer.

CASE STUDY 1: MOLDED GEAR PUMP PRESSURE BALANCE SEALS

CHALLENGE - The RP group needed to develop a technique for prototyping an injection molded pressure balance seal. Traditionally, seals are prototyped using production molds made of tool steel to withstand the high temperatures and pressures of the injection molding process. Steel molds typically require four to six weeks and $6,000 - $12,000 to complete. The first piece out of the production mold is called a "prototype". The prototype pressure balance seal needs to be tested on a dynamometer to ensure that proper sealing occurs at different pressures and oil temperatures. For an acceptable prototype, not only is a dimensionally accurate piece needed, but it also needs to be made out of a material having properties similar to the production material in terms of pliability, resistance to extrusion, etc. Castable two-part polyurethane in various durometers are available that can be poured into a mold at room temperature and pressure. These molds can be quickly and easily machined in plastic utilizing our 3D modeling and CNC machining capabilities. The specific challenge of this project was to find a castable polyurethane that would emulate the production material for testing purposes and deliver a prototype seal as quickly as possible.

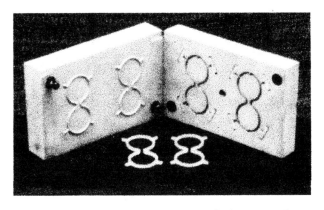

Figure 2: Prototype pressure balance seals

PROCESS - A 3D surface model of the seal was constructed from an engineering drawing. Parting lines, draft angles, and locations for pins and bushings were determined and tool paths were calculated and visually verified at the workstation. The tool paths were then downloaded to a three-axis CNC machining center. The molds were made from an easily machined, stable, high-density tooling board. The molds were sanded and waxed in preparation for the pour. A two-part polyurethane was accurately measured and mixed together by hand and placed in a vacuum chamber for degassing. The mixture was then carefully poured into one half of the mold. Aligned by the pins and bushings, the two mold halves were gently clamped together. After a 24 hour cure time, the seals were demolded.

RESULTS - The modeling, machining, and molding of the seals was completed in two days. The prototype seals were tested for approximately 45 minutes on a dynamometer cycling through various flow rates, hydraulic pressures, and oil temperatures. The seals successfully endured the harsh testing environment and verified the seal design and geometry without spending the time and money for metal tooling. After the first seal was tested, the project engineers realized that a softer material would better seal the test plate. The same molds were used to quickly mold another seal using a polyurethane of a different durometer. The engineers gained insight into how different material properties affect gear pump performance and durability. In addition to material changes, geometry iterations can be tried without the added cost and time penalties associated with production tooling modifications.

CASE STUDY 2: FIT AND FORM PUMP MODEL

CHALLENGE - A pump design requiring a new flange and body castings needed to be supplied within four weeks for an underbody clearance check. Usually this would require at least twelve weeks for the design and delivery of costly prototype castings or the lead time and expense of metal "hog-outs" just to verify the pump geometry.

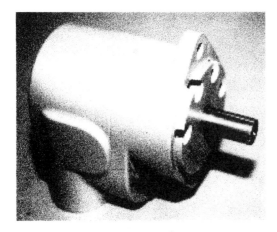

Figure 3: Plastic prototype pump

PROCESS - In order to satisfy the short lead time, it was decided that the new pump would be designed on-the-fly in a paperless process by the rapid prototyping department. The design specifications, clearance envelope, and drawings of

similar pump designs were given to the RP designer to use as a starting point. The RP designer used the initial design information to generate a 3D solid model of the new pump components. Solid models of the new body and flange were created and reviewed in three days. Toolpaths were calculated and verified at the workstation and plastic models of both components were machined to exact dimensions.

RESULTS - In less than one week the project went from initial design specifications to an assembly model in hand that was taken to the customer to verify the design and fit. All design modifications were conveyed "over the shoulder" to the RP designer eliminating the need for detailed component drawings. The major advantage of the paperless, design on-the-fly process, is the short time required for new product design and verification.

CASE STUDY 3: PROTOTYPE TOOLING FOR SAND CASTING

CHALLENGE - An order for a prototype pump body casting needed to be completed in six weeks. The normal design process for a new casting required twelve weeks for delivery and an additional two weeks for machining, assembly, and testing. Since prototype requirements had to be satisfied, the normal fourteen week lead time was unacceptable. Rapid prototyping was required to shorten the lead time to satisfy the customer.

Figure 4: CAD model of pump

PROCESS - In order to satisfy the immediate need, it was decided that the RP group would make the tooling for the new casting. Since a 2D drawing was already available and reviewed by the foundry for core box parting lines, core print locations and draft angles the RP engineer had only to develop a fully defined surface model of the tooling. Toolpaths were calculated and verified on the workstation and sent to the machine tool through a PC connection. By changing one line of code in the toolpath file it is possible to have the machine tool controller scale each X,Y, and Z coordinate by a known shrinkage value supplied by the foundry. The pattern and core boxes were machined from a high density tooling board achieving surface finish tolerance of 0.001" to insure that minimal hand sanding and bench time would be required.

RESULTS - A total of six days were required to complete the necessary prototype casting pattern and core boxes. The foundry delivered castings within two weeks and the prototype pump order was shipped to the customer within the six week requirement.

SUMMARY

These rapid prototyping examples are only a few of the projects which have proven to be successful in reducing cost and shortening development time of hydraulic fluid power prototype components. Hydraulic joystick grips and boots, and directional control valve castings have also been successfully made with the rapid prototyping technology. Our process is not state-of-the-art. It does not utilize high energy lasers or photo polymer chemistry to build parts like the cutting edge rapid prototyping systems currently on the market today. It does, however, utilize cost effective, proven machinery which is dependable and accurate. With little maintenance, beyond adding lubrication, the CNC machine is capable of producing thousands of prototype parts and tools from plastics, steel, and aluminum. The rapid prototyping industry has made great strides in the last few years to improve accuracy, speed, and reliability. The combination of current machine tool technology with advanced 3D CAD modeling capabilities will continue to perform a major role in meeting the future prototype requirements of our customers.

Programmable Logic Control of Gerotor Motors Incorporating Speed Sensing and Dual Cross Port Valving

Gregory A. Schmidt
White Hydraulics, Inc.

ABSTRACT

The use of Hall Effect sensing and crossport valving in gerotor motors has opened the possibilities for electronic feedback control in hydraulic applications. By utilizing a "zero-speed" Hall Effect speed sensor and a motorized bypass control valve, the flow into the motor can be adjusted to obtain a desired speed or rate. Incorporation of PLC (programmable logic control) hardware will enable applications to respond to variations in load. The advantages of the speed sensor and control valve contained in the motor along with a small, light weight PLC, lends itself to a viable option in many control applications.

INTRODUCTION

With the increasing use of hydraulics in mobile and control applications, a need for unique and novel design strategies are being sought. Finding a feasible way of monitoring hydraulic motor characteristics has been difficult, and maintaining those characteristics during operation almost impossible. Any control changes that must be given are usually supplied to the hydraulic pump or flow orifice and not to the actuator. This frequently causes sluggish, uncontrollable, inappropriate, or unwanted operation. These obstacles provided the opportunity to develop a means of allowing "automatic" control of a hydraulic system directly at the actuator, in this case a hydraulic gerotor motor.

The development of four key products provided innovation of self-controlling and automated hydraulic systems. First, the evolution of using Hall Effect speed sensing in gerotor motors. This allows sensing of rotational speed (RPM) and direction, when used with 2 sensors (quadrature). Second, a dual cross port valve cavity in the motor junctions the 'A' and 'B' port internally without the use of a manifold block. This cavity commonly is useful for a bi-directional relief valve or dump valve to bypass the fluid from the motor which will cause it to stop without turning off the flow. Third, a motorized bypass control valve provides electric remote flow control. Finally, a lightweight and inexpensive programmable logic controller (PLC) collects the information from sensors, processes the information, and returns the proper output to controlling devices. By combining these products, it is possible to have economical, lightweight programmable logic control at the hydraulic motor or actuator.

THEORY

HALL EFFECT SPEED SENSOR - The Hall Effect sensor supplies a digital square wave pulse train as magnetic poles pass under the sensor. Every time a south magnetic pole passes, a signal is generated, and when a north pole passes the signal goes off. The device is "zero-speed" in that the signals are turned on or off, regardless of speed. With a 120-pole ring (60 north poles and 60 south poles) 60 pulses per revolution are provided.

MOTORIZED BYPASS CONTROL VALVE - The control valve is installed in a standard 10-2 cavity. To adjust flow control, the valve uses a 12-volt DC motor to open or close a rotary spool valve. Flow adjustment is changed by reversing the polarity of the signal going to the valve. The 12 volt DC signal is only applied to change the flow setting. If no signal is given, the valve maintains its flow setting. Figure 1 shows the specifications for the valve.

Figure 2 shows a hydraulic motor with a motorized control valve and a speed sensor.

APPLICATIONS

The first applications for speed sensors on gerotor motors were to monitor flows on test stands. This

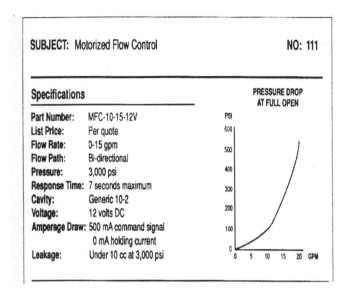

Specifications		PRESSURE DROP AT FULL OPEN
SUBJECT: Motorized Flow Control		**NO:** 111

SUBJECT: Motorized Flow Control   NO: 111

Specifications

Part Number:	MFC-10-15-12V
List Price:	Per quote
Flow Rate:	0-15 gpm
Flow Path:	Bi-directional
Pressure:	3,000 psi
Response Time:	7 seconds maximum
Cavity:	Generic 10-2
Voltage:	12 volts DC
Amperage Draw:	500 mA command signal
	0 mA holding current
Leakage:	Under 10 cc at 3,000 psi

Figure 1: Motorized valve specifications

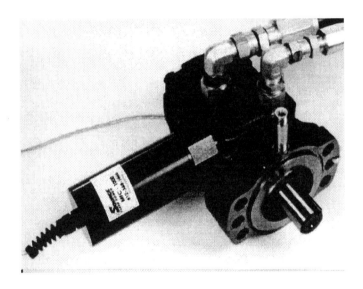

Figure 2: Hydraulic motor with Hall Effect speed sensor and motorized control valve

provides accurate flow information at wider speed ranges as compared to in-line turbine flow meters at low flows. Using a PLC with the motorized bypass control valve and a speed sensor, tests can be performed at constant speeds and loading. It can be difficult to keep parameters constant during performance or endurance testing, especially as fluid temperature increases. As the temperature increases, the fluid thins, which usually decreases the efficiency of the system. Pressure falls as does flow. By testing motors against a load pump (or motor) equipped with a speed sensor and motorized valve in close loop, constant speed and loading can be

successfully achieved, independent of input fluctuations.

As awareness of environmental concerns heighten, accurate control of salt, sand, and liquid spreaders and de-icers through closed loop systems will prevail, Figure 3. The signal and voltage range produced by the Hall Effect sensor works well with many of the smaller PLC's that are available. Published reports state that the cost of the salt spread is a small fraction of the cost of repairing road surfaces ($700-$800 for repairs for every ton of salt, $30 per ton). Accurate PLC control of spreading rates reduces damage to paved surfaces and runoff while optimizing surface traction.

Figure 3: Salt trucks, Milwaukee Highway Dept.

This PLC configuration has been successfully used in lime spreading systems (see Figure 4). Uniform spreading of lime can be achieved for virtually any condition

Figure 4: Lime spreader

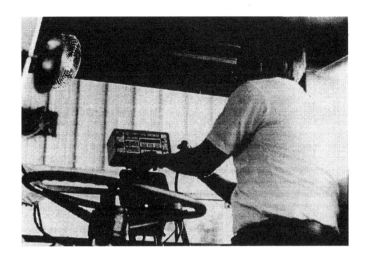

Figure 5: Spreader Adjustment

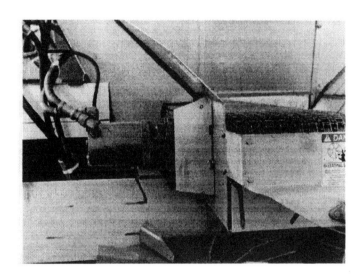

Figure 6: Hydraulic motor driving conveyor

variations or truck speed. A familiar problem with spreading lime is that as the load of lime lessens, the conveyor speed picks up and too much lime is spread, resulting in uneven spread and possibly underestimating the amount of lime required for an area. The arrangement of using a PLC with the motorized valve assures a constant spread. Figure 5 shows how spread rate can be adjusted which changes the flow to the hydraulic motor in Figure 6. This system may also be connected with truck speed. As the speed of the lime spreading truck varies while dispensing lime, conveyor speed can be instantly corrected.

Positioning systems are an ideal application for speed controlled gerotor motors. A robotic melon harvester built and tested by the University of Kentucky, utilizes a speed sensor controlled motor for positioning. The evaluation

confirmed that the position of the robot could be located to within a single centimeter. The U.S. Army is now developing a prototype version of an autonomous exterior physical security robot for the Mobile Detection, Assessment, and Response System (MDARS) program, Figure 7. These systems will patrol around warehouses, industrial parks, and ammunition supply depots. Advanced technology will be incorporated in the areas of autonomous command, control and navigating; video compression, image processing; and spread spectrum, RF low-band communications. Hydraulic speed sensing motors were furnished for the prototype RST machine.

Figure 7: MDARS vehicle concept

These are only a few of possible applications that can be designed using a motorized flow control valve and a PLC incorporated into a hydraulic motor.

PROGRAMMING SAMPLE

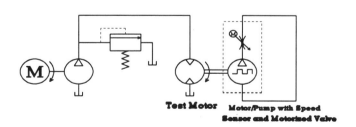

Figure 8: Hydraulic motor test schematic

This is a simple application that provides constant loading for hydraulic testing purposes. As discussed earlier, this will allow constant speed and load to a test hydraulic motor that drives another motor (as a pump) containing the motorized valve and PLC operated in closed loop. See figure 8 for schematic and Figure 9 for control circuit. Figure 10 shows the sample Ladder Diagram and the program list. The pulses fed into the input relay, No. 0000, are counted. If the number of pulses within a specified time is less than the predetermined value, then the output relay, No. 0500, is activated; if the number of pulses is more than the predetermined value, then output relay, No. 0501, is activated. These outputs are tied into the motorized bypass control valve, which opens or closes to adjust the

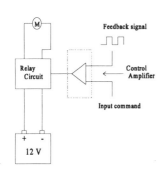

Figure 9: Control circuit

amount of fluid entering the motor. Thus, running speed is precisely regulated. To vary the speed, the specified time can be changed, TMR #14, via an analog timer. To increase speed, the timer would be lessened; to decrease speed the timer would be increased.

SUMMARY

In conclusion, by combining a Hall Effect speed sensor and a motorized control valve with a PLC, control of a hydraulic motor or actuator is feasible. This combination of unique products replaces heavy, expensive, and cumbersome alternatives. Another unique advantage is having the Speed Sensor and Control Valve located at the source of output, and not "upstream" from point of actuation.

REFERENCE

Source Fluid Power, Sales Bulletin #111, February 1994

Keyence, Pocket-sized Programmable Logic
        Controller, 1991

White Hydraulics, Newsletter #69, April 1995

Dowling, James M., "Gerotor Motor Hall Effect Speed
        Sensing", #911599, American Society of Agricultural
        Engineers, St. Joseph, MI, 1991

0000	LDB	CTR00	
0001	LD	TMR14	
0002	LD	2008	
0003	SFT	1000	1000
0004	LDB	CTR01	
0005	LD	TMR14	
0006	LD	2008	
0007	SFT	1001	1001
0008	LD	0000	
0009	LD	TMR14	
0010	OR	2008	
0011	CTR	00	#0010
0012	LD	0000	
0013	LD	TMR14	
0014	OR	2008	
0015	CTR	01	#0015
0016	LD	1000	
0017	OUT	0500	
0018	LD	1001	
0019	OUT	0501	
0020	LDB	TMR14	
0021	TMRH 14		#9999
0022	END		

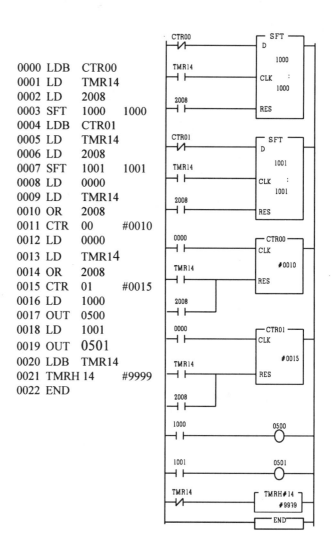

Figure 10: Ladder diagram

Hydraulic System of John Deere 8000 Series Row Crop Tractors

Gregory E. Sparks
John Deere Product Engineering Center

ABSTRACT

The John Deere 8000 Series agricultural tractors, introduced in August 1994 represent a dramatic change in visibility, maneuverability, power, and control unprecedented in tractors this size. It's basic chassis configuration was even granted a U.S. patent. The hydraulic system is significantly different than the previous John Deere models they replace, but similar to changes started by the 7000 Series tractor design. The new system adopted the closed center pressure-flow compensated design for tractor/implement control.

Totally new components are used for all systems including pumps, brakes, electro-hydraulic remote valves and hitch, breakaway couplers, steering, charge circuit, and filtration. The new system provides performance improvements in several areas and an exclusive CommandArm$_{TM}$ and unique TouchSet hydraulic control panel for unmatched operator convenience and control. This paper describes the hydraulic system structure, components, and some of the engineering considerations dealt with in the development process.

INTRODUCTION

The new line of agricultural tractors introduced by John Deere in 1994 ranging from 119 to 168 PTO kilowatts (kW) was refined by countless hours of customer research, competitive tractor assessment, unprecedented supplier involvement, and extensive testing to meet the needs of our customer. This family of tractor models include the: 8100 with 119 PTO kW, 8200 with 134 PTO kW, 8300 with 149 PTO kW and the 8400 with 168 PTO kW. All models are manufactured in Waterloo, Iowa.

In addition to the objectives covering safety, performance, reliability, and cost, there was an objective to get our preferred suppliers involved from the onset of the design to optimize engineering and manufacturing investments. The 7000 Series models introduced two years earlier helped establish our initial supplier base and also defined the objectives and criteria to be used in judging program performance. General categories used in the assessment were performance/reliability of the design, cost, quality, engineering responsiveness, production flexibility and delivery, product fit with existing business, and financial stability.

Throughout the design and development program Multifunction Engineering Teams were utilized. Every tractor functional group had representation from their respective engineering and manufacturing disciplines. Typical representation included internal Deere personnel ranging from Purchasing, Vehicle and Systems Engineering, Quality Engineering, Product Evaluation and Marketing. Outside representatives came from Supplier's Manufacturing and Design Engineering groups. Meetings and design reviews were conducted when needed and held where it was most productive, at Deere engineering/manufacturing facilities or at suppliers engineering/manufacturing facilities.

This paper will cover the 8000 Series tractor hydraulic system.

SYSTEM CONFIGURATION

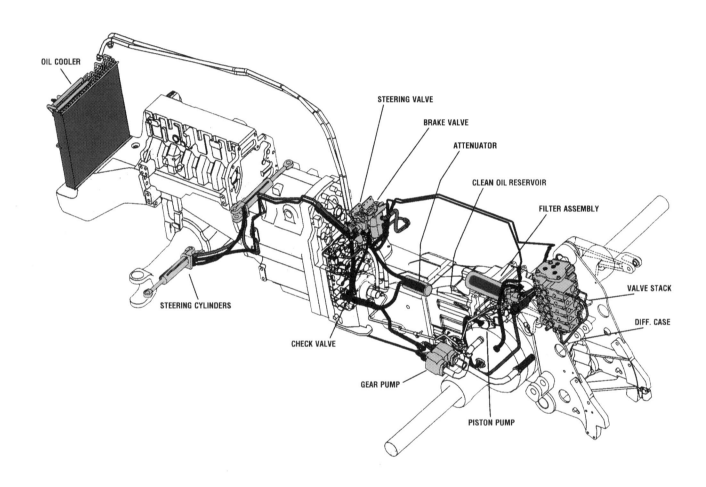

FIGURE 1- HYDRAULIC COMPONENT (PICTORIAL) LAYOUT

Due to the unique chassis changes required to enhance visibility, maneuverability, and increased fuel carrying capacity of these large frame tractors, the hydraulic layout is significantly changed from the 60 Series tractor.

SYSTEM DESCRIPTION AND OPERATION

PUMP AND DRIVE LOCATION- Hydraulic pressure and flow is provided by a 45 ml/rev axial piston pump for remote valve/ implement utilization and a tandem gear pump. One section of the gear pump is used for low pressure filtration and axial pump charging and the other section for tractor control (steering/brake/ cooling/and transmission control). Their displacements are 40 ml/rev and 31 ml/rev, respectively. The hydraulic pumps were located low and toward the rear of the tractor. The pumps are driven by a hypoid gear set, with a gear ratio approximately 120% of engine speed. The driven shaft provides the common drive speed for both the tandem gear and piston pump.

PARTIAL CHARGE OIL RESERVOIR- The 27 liter clean oil reservoir (COR) enables the continued use of a partial charge system by placing it above the pump inlets and ensures a positive pressure head to the axial piston pump. Utilizing a smaller displacement charge pump minimizes power loss and cost without sacrificing performance. The differential/axle assembly serves as a common reservoir and holds approximately 65 liters of oil.

FILTRATION AND GEAR LUBE- One filter assembly effectively maintains system cleanliness levels at or below an ISO 20/15 level and was manifold mounted directly to the COR to minimize plumbing. The large 127mm base spin-on filter replaces two cartridge/cover assemblies. The COR also provides filtered oil for the tractor control gear pump. Sizing differences between the charge and control pumps enabled direct oil lubrication of the differential ring gear/pinion interface through the overflow passage out the top of the COR.

COMPONENTS OUTSIDE OF CAB- Both steering and brake valves were located outside and under the cab to eliminate hydraulic oil presence in the cab. The steering unit is mounted to the cab using rubber isolation bushings and the brake valve is rigid mounted, to ensure good mechanical efficiency for single stroke manual braking, of these large tractors. An attenuator was used to provide hydraulic capacitance to suppress hydraulic gear pump ripple noise transmission.

PTO/PUMP DRIVE AND BRAKE LUBE OIL- A scavenge gear pump in the transmission sump returns oil from the front of the tractor rearward through the lower rotating pump drive shaft providing cooled lubrication oil for pump drive gears, PTO clutch, and both left hand and right hand wet brake discs. It has a 102 ml/rev displacement and rotates on the PTO drive shaft at 45% of engine speed.

PLUMBING- To minimize plumbing, components are manifold mounted where possible. Steel lines are used to connect components where relative motion is negligible and hoses are used where components see relative motion or to reduce noise transmission. Metric flat face o-ring connectors and fittings are used throughout the hydraulic system. Total wet hydraulic oil volume is approximately 140 liters.

SYSTEM DESCRIPTION AND OPERATION

The total tractor hydraulic system can be divided into three circuits for convenience in describing function and operation. These circuits are: (1) Low Pressure (Filtration and Charge) Circuit, (2) Tractor Control Circuit, and (3) High Pressure (Implement Oil) Circuit.

LOW PRESSURE CIRCUIT- The low pressure (filtration and charge) circuit includes the components operating at a maximum pressure of 415 kPa. (This elevated pressure occurs during maximum piston pump flow demands and typically runs near the 20 kPa level.) The differential case and axle assembly serve as the main fluid reservoir for both the transmission and hydraulic system. The charge section of the tandem gear pump draws fluid through a suction screen from this sump.

The charge pump flow is filtered and directed to either the piston pump inlet or on to the clean oil reservoir (COR). Large 38mm diameter inlet tubes provide sufficient flow paths to the pump, especially during cold weather operation.

When the piston pump is at standby (no flow demand), the oil is routed to the COR which is used by the tractor control pump, excess oil flow is used to lube the differential ring gear/pinion interface located in the differential case.

When the axial piston pump is flowing oil, the oil normally returns to the filter assembly to help maintain a positive inlet pressure condition to the axial pump. Dual acting cylinder applications return oil to the filter assembly which helps maintain a positive inlet pressure charge of 20 to 35 kPa at the piston pump inlet. Single acting cylinder applications draw makeup oil from the COR in a naturally aspirated mode at 35 kPa maximum vacuum.

Located between the filter and COR assemblies is a restrictor orifice that ensures a positive inlet pressure on the axial piston pump for the majority of pump operating conditions.

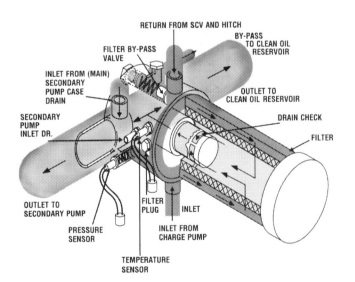

FIGURE 2- FILTER ASSEMBLY CROSS-SECTION

When differential case oil level is at the full mark, up to 30 liters of oil can be drawn from the tractor for use in raising large dump wagons. If the COR is pulled down , a float switch detects low reservoir conditions and provides an electrical signal to shut off "commanded" electro-hydraulic (EH) remote valve and hitch raise signals. Oil can be returned by commanding the opposite EH remote valve signal or by lowering the hitch. Normal function is regained once the reservoir level is refilled.

A single spin-on type oil filter is used for both hydraulic and transmission circuits. The filter head contains a differential pressure switch for advanced warning of a potential upcoming bypass condition, a sensor for monitoring hydraulic oil temperature, and an anti-drain back valve. The filter bypasses downstream and the filter bypass switch provides a warning signal which is processed by the Chassis Control Unit (CCU) to determine if the warning light should be illuminated in the cab. The system utilizes one 127mm filter head assembly with a filter capacity of 100 grams per ISO 4572 and filter efficiency of $\beta_{10}= 9.9$ TWA. Typical service life is 750 hours or once a year. The anti-drain

valve enables the filter to be removed without draining the COR.

Several benefits were attained:
1. A partial charge/NA system that reduces system cost without sacrificing performance. Inlet tubes of ϕ 38mm inlet tubes were necessary to maintain satisfactory cold oil performance.
2. Single filter assembly that maintains desired ISO system filtration needs and desired yearly service interval.
3. Adequate oil take-off for single acting dump wagons with built in safe guard logic to minimize and even eliminate system component damage due to inadvertent or accidental implement hose breakage.

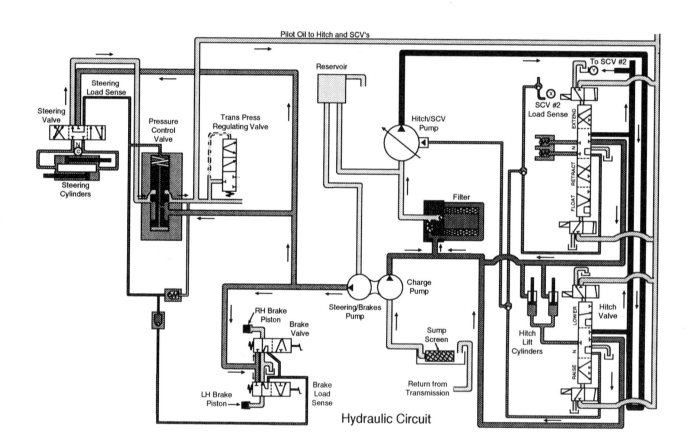

FIGURE 3- 8000 SERIES TRACTOR SIMPLIFIED HYDRAULIC SYSTEM SCHEMATIC

TRACTOR CONTROL CIRCUIT- The tractor control circuit consists of the components used to provide steering, service brakes, system cooling, system lubrication and transmission control functions.

The clean oil reservoir (COR) serves as this circuits source of oil. The tractor control pump section of the tandem gear pump draws oil from the COR through a large diameter suction tube and directs the flow of oil to

the pressure control valve which is manifold mounted directly to the steering valve.

A supply line to the service brake valve, and a connection to the ground drive logic valve, in the rear transmission cover, is T'd off of this circuit prior to the pressure control valve.

The pressure control valve located at the steering valve in this circuit is basically a sequence valve. The

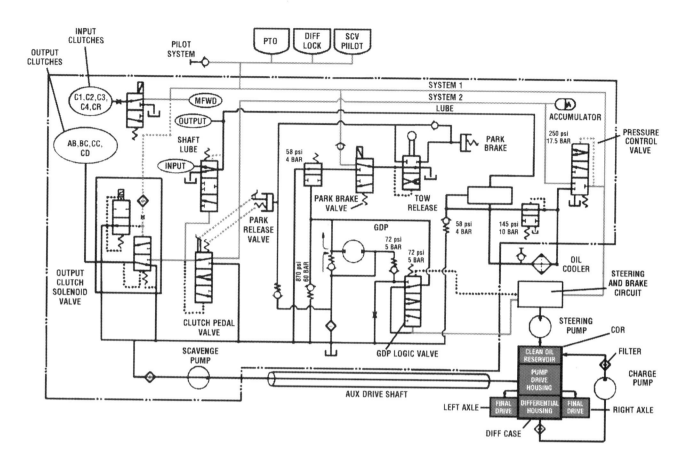

FIGURE 4- SIMPLIFIED TRANSMISSION CONTROL CIRCUIT SCHEMATIC

reference line to the sequence valve spring chamber is the load sense signal from steering or brakes. This enables the high pressure side plate gear pumps pressure to be controlled from approximately 3,000 kPa at standby to 20,800 kPa at stall. Static margin is controlled by the compression spring at 700 kPa and dynamic signal boost to 1,100 kPa is obtained by the orifices within the combined steering and pressure control valve assembly.

Excess flow through the pressure control valve is routed to the transmission control circuit where an additional sequence valve maintains transmission control pressure of 1,750 kPa. The control oil is then utilized for transmission shifting, MFWD, PTO, differential lock, and remote valve pilot control.

Excess flow through the transmission control (sequence) valve is routed to the combined hydraulic/transmission oil cooler. Approximately 1.0 l/s oil flow through the hydraulic oil cooler maintains sufficient system oil temperatures under the worst field conditions. A cooler bypass valve in the transmission valve cover opens at 1200 kPa in cold weather conditions.

Cooled hydraulic oil returning from the oil cooler is used for transmission lubrication. The transmission shafts are drilled to strategically utilize the oil flow and

minimize windage loss. The lube pressure is limited to 200 kPa by an additional relief valve.

The oil is retained in the front transmission sump for ground drive pump (GDP) usage and then overflows to the scavenge pump sump to be picked up by the scavenge gear pump and moved through the lower pump drive shaft. The pump is sized to 125% of transmission control pump displacement to provide air pressurization in moving the oil rearward. The pump oultet air/oil pressure is less than 100 kPa to minimize air entrainment. This oil is used to provide cooling and lubrication of the pump drive gears and PTO clutch in the pump drive housing mounted to the differential case front face. The PTO clutch housing oil compartment is air tight, and located low on the tractor. Oil has to be lifted approximately 750mm to provide cooling and lubrication oil to both left hand and right hand brake discs. Air pressure acting on the oil in the PTO clutch housing helps lift the oil out two identical φ22 mm oil tubes. The pump drive chamber air pressure is approximately 30 kPa with warm oil (20 cSt) and can reach 400 kPa in extreme cold temperatures (20,000 cSt). Oil in both outer axle assemblies reach their spill dams near axle centerline and then overflow back into the center differential case sump to complete the oil flow process.

Benefits of the circuit:

1. Oil flow noise reduction was obtained with the higher back pressure acting on the steering control valve.
2. Better utilization of the gear pump circuit by combining both low and high pressure functions through use of load sense methodology.
3. Separation of the steering and brakes from the piston pump circuit to minimize end of field system interactions and flow loss.

HIGH PRESSURE CIRCUIT- The high pressure circuit provides hydraulic power to implement hitch and remote valve functions. The pressure flow compensated axial pump provides an oil source to only the hitch and remote valves.

The high pressure circuit can be categorized by commonly used terms of: "closed center", "PFC" (Pressure and Flow Compensated), "POD" (Pressure or Power on Demand). The nine piston axial pump has integral control valves for high pressure limit and load sensing flow control. The pressure limit control is set at 20,000 kPa and the load sensing flow control provides a margin, or differential control pressure, of 2,000 kPa. All 8000 series tractors utilize a 45 ml/rev pump and 40 ml/rev charge pump. The piston pump is mounted on the right hand side of the pump drive housing driven via a hypoid gear set at 120% of engine speed. Flow capacity is approximately 1.9 l/s at an engine rated speed of 2200 rpm.

The major benefits of this circuit:

1. Separate independant oil circuit for implement utilization.
2. Simple circuit with good stability and ease of trouble shooting

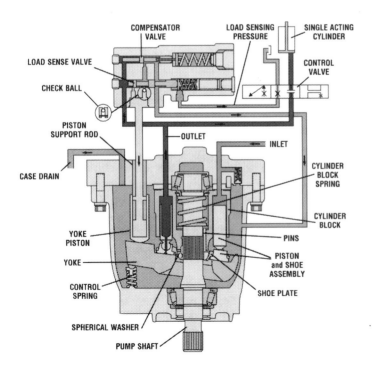

FIGURE 5- AXIAL PISTON PUMP, 45 ML/REV

SERVICE BRAKES- Wet disc brakes are activated by a power brake valve mounted on the outside surface of the operator enclosure. The power brake valve, approximately 20% larger than the 7000 Series tractor valve, provides independent braking for left and right axles for steering assist, but has an equalizer function to ensure equal pressures are provided to both brakes during simultaneous braking.

The brake valve functions as a proportional pressure reducing valve with output pressure being proportional to input force. An actuation spring is positioned in series with the force transmitting elements of the valve to provide increased pedal travel for better modulation characteristics. The output pressure sensing area of the brake valve, brake pedal ratio, and axle brake piston area were the variables adjusted to meet 4.5 m/sec2 peak deceleration at maximum homologated vehicle weight with no more than 800 Newtons pedal force, (a requirement per German STVZO sec. 41). Brake facing friction, final drive ratio and rolling radius were the other important parameters in the design analysis. Other documents governing the design include ASAE S365.2 and EEC OJ No. L122,8.5.76.

The higher pressure of the left or right brake is selected by a shuttle check valve within the brake valve housing and directed via a load sense shutoff valve to the pressure control valve mounted to the steering valve. The pressure control valve assures that the power brake valve operates at a minimum pressure differential of 1,000 kPa.

Although the typical function of the brake valve is to meter pressure and flow from the gear pump to provide power brakes, a second less, but still important function is to provide manual braking capability if power is lost.

The power brake valve on the 60 Series agriculture tractors of this size utilized an accumulator to store energy for this purpose. That concept worked well with the closed center pressure compensated system used on those tractors, since pressure was maintained near the rated system pressure thus keeping the accumulator fully charged. The 7000 and 8000 Series system, due to its ability to operate at low pressure much of the time would require a complex and costly charging valve to keep an accumulator charged.

The 8000 Series brake valve was a spin off of the 7000 Series valve developed with capability to operate as a dual stage manual valve. In the absence of supply pressure, 1) the load sense shutoff valve closes, eliminating this leakage path out the brake circuit and 2) the manual piston dump valve shifts to block prefill piston leak path to sump. The dump valve was added to the 8000 Series design due to the larger prefill oil requirements of our large axles and the need to dump this oil in the power mode during cold oil conditions to minimize a heavy pedal "feel". When the operator applies force to the brake pedals, the initial portion of the stroke forces oil from the large prefill piston chamber, through a check valve in the power metering spool to the brake lines and axle brake pistons. When the axle brake piston moves through the retraction distance and establishes contact between the piston, friction disc and backing plate, pressure will begin to increase. At approximately 1,100 kPa, the prefill chamber relief valve opens, causing the pressure to drop and the check valve to close. At this point, a transition is made such that input force now acts to create additional flow and pressure via the smaller diameter spool acting as a piston. In this way the operator can generate the higher

pressures required to stop the tractor without excessive pedal effort or travel.

An optional hydraulic trailer brake with coupler per French Standard NF-U-16-006 (ISO 5676) is available. This brake valve receives the load sense signal from the tractor brake valve and provides trailer brake pressure from the piston pump circuit proportional to tractor brake pressure up to a maximum of 15,000 kPa.

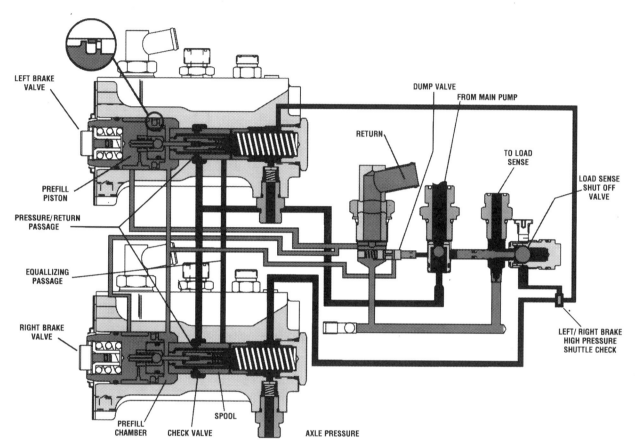

FIGURE 6- 8000 SERIES TRACTOR BRAKE VALVE

STEERING- The position responsive hydrostatic power steering system consists of two double acting cylinders mounted on the front axle and a steering control unit mounted on the outside surface of the operator enclosure with input via the steering wheel. The steering control unit is a non load reactive flow metering type with integral gerotor hand metering pump. Metering pump displacements of 185 ml/rev for two wheel drive and 231 ml/rev for tractors with mechanical front wheel drive (MFWD), are used to provide 3.9 and 3.75 steering wheel turns lock to lock, respectively.

The steering circuit utilizes a dynamic load sense signal provided by the integrally mounted pressure control valve (PCV). The PCV was integrally mounted to improve system performance, reduce plumbing complexity and minimize cost. The dynamic load sense signal (signal flow goes to load) tends to improve response, especially at cold temperatures.

The pressure control maintains a 1,100 kPa pressure differential between steering unit supply and load sense. Steering supply pressure is limited to 20,800 kPa by a relief valve in the steering control valve signal passage.

Steering system torque measured at the axle kingpin was increased by 20% over the 60 Series tractor. The increase was a result of increase hydraulic force (pressure and area) and a decrease in caster angle to 5°.

Auxiliary, or "dead engine" steering on either two wheel drive (2WD) or MFWD tractors is provided by an auxiliary power source. "Dead engine" steering on the 2WD tractor could be provided by the operator input power to the steering wheel to meet EEC/ECE 75/321. However, this requirement can not be met on the larger MFWD tractors, due to increased front weight capacity and different steering geometry. In addition tractors sold in Germany with road speeds up to 40 kph must meet

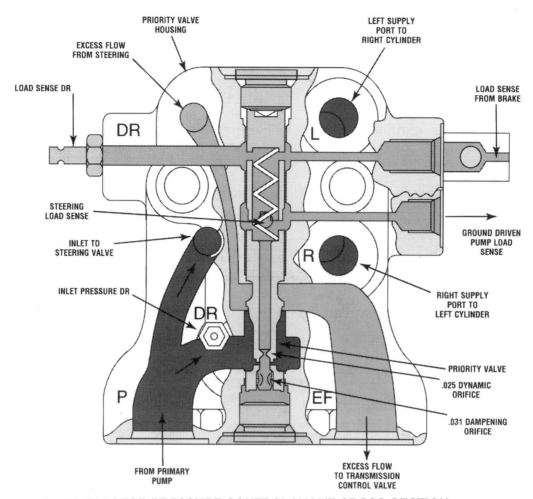

FIGURE 7- 8000 SERIES TRACTOR PRESSURE CONTROL VALVE CROSS-SECTION

the more stringent STVZO German road regulations for auxiliary steering. In order to meet this requirement, an auxiliary power source is required. This need has been met on all 8000 Series tractors by use of a ground driven auxiliary steering pump (GDP). This is a fixed displacement gear pump driven by the differential drive shaft to the rear wheels. The pump is integral to the transmission and located in the differential drive shaft (DDS) quill. The integral control valve which recirculates flow internally at low pressure drop when the normal power steering is functioning is located in the rear transmission valve cover. The low pressure signal from the pump is also used to shuttle another logic valve to block park brake sump passage. The park brake on these tractors is a spring applied, pressure off design. An EH on/off valve enables this function. Therefore, all 8000 Series tractors are equipped with the GDP, whether its a 2WD or MFWD version, to eliminate the park brake from engaging at speeds above 2 kph due to an electrical or mechanical failure of the park brake solenoid valve. When the differential pressure to the steering unit drops from the normal 1,100 kPa to approximately 500 kPa, the control valve diverts flow to the steering unit to provide power for steering. An internal relief valve in the pump outlet passage limits pressure from this pump to approximately 6,000 kPa.

This reduced steering capacity is adequate to meet the STVZO requirement, but the decrease in performance will be noticed by the operator so corrective action will be sought.

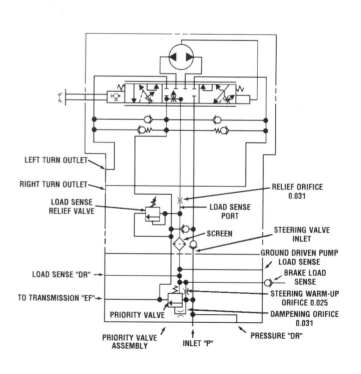

FIGURE 8- 8000 SERIES STEERING SCHEMATIC

98

ELECTRO-HYDRAULIC VALVE STACK- Valves for remote function control and 3-point hitch control are sectional valves mounted in a vertical stack outside the cab, behind the operator. The stack is manifold mounted on top of the hitch frame with all hydraulic connections made through internal machine passages. The stack contains the necessary shuttle check circuit to communicate the highest load pressure to the pump load sense control valve.

Three electro-hydraulic (EH) remote valves with EH hitch valve is the standard factory equipment, with a fourth EH remote valve as a factory option. A fifth section can be added as a field installed option. A no hitch stack can be obtained as well.

Auxiliary load sensed power beyond capability is available in either a factory or field installed kit. The load sense signal provides a bleed off orifice to return for those auxiliary valve packages that do not bleed load sense pressure to return in neutral.

FIGURE 9- 8000 SERIES TRACTOR EH VALVE STACK

REMOTE FUNCTION CONTROL SYSTEM

Since one of the major objectives of the program was to provide 21st Century Technology today, electronically controlled remote valves was a necessity.

Design teams were established to identify a user friendly control system that would allow easy changes in control actuation, duration of establish flow, and control of flow rate over it's entire flow range, by the touch of a finger.

This led Deere engineers to the revolutionary control of tractor hydraulics from your tractor seat by creation of the Deere and TouchSet Hydraulic Control Panel.

The remote valves are electro-hydraulically controlled through use of: operator controlled input potentiometers or switches, a microprocessor based Selective Control Unit (SCU), and a proportional solenoid.

Operator controls consists of:
1. Hydraulic control levers on the CommandArm that provide control with extent, retract, float, and timed detent position. The lever's also provide a proportional flow signal, the farther you push, the more oil flow. One lever is provided for each remote valve function.
2. One common TouchSet Hydraulic control panel. Roman numeral touch switches activate a timed detent control knob, to adjust duration of flow on time, and additional knob to set the relative oil flow rate.

Remote valve "timed" detent can be adjusted on the go from your opeartor seat between one to nineteen seconds. The clock can also be set at "0" seconds to denote "no detent function" (ex: loader operation) and "C" continuous detent operation (ex: motor operation).

Remote valve proportional flow adjustment can also be done on the go, from inside the cab, from 0.01 l/s to 1.9 l/s through 100 distinct settings of the flow control knob.

The SCU in addition to being connected to the valve control components, is linked to the RH front post tachometer module via an interface communication link with the chassis control unit (CCU). The tachometer module serves as a window or monitor screen to set and check the values which must be stored in the SCU for proper remote valve control. The values are stored by operating the remote valve system in a "calibration" mode. Valve "deadband" value has to be determined and stored to make the valve responsive to commanded operator input. The SCU, via the tachometer, leads the factory assembler or dealer service technician through a series of steps which determine proper calibrations for each valve solenoid. Once calibrated, the SCU is returned to normal operating mode where it receives signals from operator controls and determines how much current to send to the extent or retract solenoid of the remote valve.

REMOTE VALVE- A clean sheet approach was taken to provide an advanced designed electro-hydraulic remote control valve. Patents were obtained on design features of this valve. All EH remote valves sections provide 4 way-3 position flow control using two electrohydraulic proportional solenoids to pilot the main spool. It directs flow from the axial piston pump to implements interfaced through the remote valve couplers, one for pressure oil out to a cylinder, and from the cylinder to return.

Each section has an inlet pressure compensator to maintain constant pressure drop on the pressure metering area regardless of implement load or pump pressure. Adequate return metering is maintained to control over running loads.

Two hydraulic pilot valves are provided to shift the main spool. These are proportional pressure reducing/relieving valves which provide a pilot pressure proportional to current in the solenoid coil. One pilot valve provides pressure to shift the main spool to extend

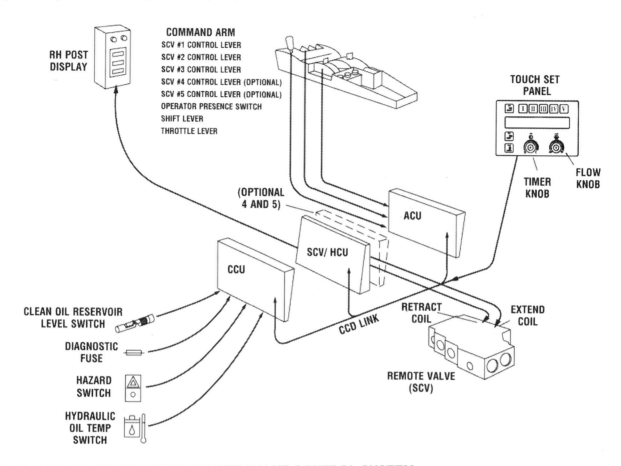

FIGURE 10- 8000 SERIES TRACTOR REMOTE VALVE CONTROL SYSTEM

position (pressurize the LH coupler passage and open the RH passage to return) and the other to shift the opposite direction to retract position (pressurize the RH coupler passage and open the LH passage to return).

Pilot operated lockouts on each work port with thermal relief valves maintain load holding leakage to less than 1.5 ml per minute. Leakage is measured at a work port to return pressure drop of 10,000 kPa and 20 cSt (20 cSt).

Float function is obtained by piloting both lockouts open and leaving the main spool in neutral. Pilot pressure also acts on a float spool valve which couples the extend and retract ports together to allow free passage of oil.

Maximum available work pressure from each remote valve at 1.2 l/s is 17,000 kPa and at 1.9 l/s is 13,800 kPa. Peak valve output power is 26 kW at 1.9 l/s. (All data measured at 20 cSt).

Remote valve flow versus command signal response is designed for both open and closed loop position control of implement cylinders and motors.

HYDRAULIC BREAKAWAY COUPLER- Hydraulic breakaway couplers for the remote control valves are integral to the remote valve housings and are located at the left side of the rear of the tractor. These couplers

accept the ISO 5675 size 12.5 male coupler tips and are located per ASAE 5366.1.

The male coupler is connected by pushing into the female coupler. When inserting a pressurized male coupler, the female coupler is pushed toward the front of the tractor. This opens a valve which vents the anti check piston, allowing the female coupler poppet to open. the male coupler can therefore be placed in position for the locking balls to engage the locking groove with very little effort. Once coupled, a centering spring returns the female coupler to it's normal position trapping the locking balls in the groove.

Pressure from operation of the remote control valve opens the lip seal and acts on the anti-check piston forcing the male coupler poppet open, allowing flow. Pressure is trapped in the anti-check piston cavity preventing flow checking or closing of the male coupler poppet under high flow conditions. Uncoupling is accomplished by pulling on the hose. This action moves the female coupler and locking collar toward the rear of the tractor. This opens the valve, releasing pressure from the anti-check piston, allowing the male coupler to close and bleeding off trapped pressure. Since the separating force between the male coupler and the receptacle is now reduced, the locking balls easily retract from the groove releasing the hose. A lever with mechanical advantage to move the female coupler is provided on the hydraulic couplers of 8000 Series model tractors to allow simultaneous release of both coupled hoses without pulling on them.

Maximum connecting pressure is 40,000 kPa. Pressure drop through both couplers in series is 900 kPa maximum at 1.5 l/s oil flow (20 cSt).

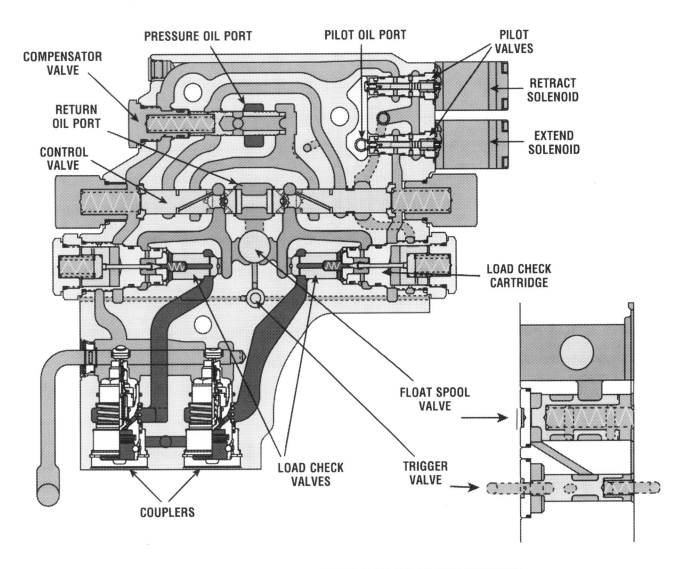

FIGURE 11- 8000 SERIES TRACTOR REMOTE VALVE W/ COUPLER CROSS-SECTION

THREE POINT HITCH POSITION AND DRAFT CONTROL SYSTEM

The 8000 Series tractors can accommodate implements with hitch categories of 3 or 3N. The 3-point hitch is electro-hydraulically controlled through use of: operator controlled input potentiometers or switches, hitch sensing potentiometer, a microprocessor based hitch control unit (HCU), and a proportional solenoid operated valve.

The hitch system is very similar to the 60 Series model EH hitch, except for location of control levers in the CommandArm and the change from two proportional two-way control valves to a single spool type three-way EH control valve.

Operator controls with potentiometers are:

1. The hitch position lever to raise and lower the hitch and set implement working depth,
2. The rate of drop knob to adjust hitch drop speed to match implement weight and operating needs,
3. The raise limit control to vary the height the hitch will raise to when the raise/lower switch is actuated,
4. The load/depth control to vary the control mode from pure position control to maximum sensitivity to draft or implement load.

Operator controls utilizing switches are:

1. The fast raise/lower switch to quickly raise the implement to the preset raise limit and lower to the working depth as set by the hitch position lever,
2. The external raise/lower switch mounted at the rear of the left fender to slowly position the hitch to aid in attaching implements.

Two hitch sensing potentiometers are used:
1. The position feedback potentiometer to provide a signal proportional to hitch position,
2. The draft (load) sensing proportional to load on the lower draft links.

The Hitch Control Unit (HCU), in addition to being connected to the hitch control components, is linked to the RH front post tachometer module via an interface communication link with the chassis control unit (CCU).

The tachometer module serves as a window or monitor screen to set and check the values which must be stored in the HCU for proper hitch control. The values are stored by operating the hitch system in a "calibration" mode. Two basic values are lift cylinder size (indicating flow requirement), and tractor model (indicating power or draft capability). The HCU, via the tachometer, leads the factory assembler or dealer service technician through a series of steps which

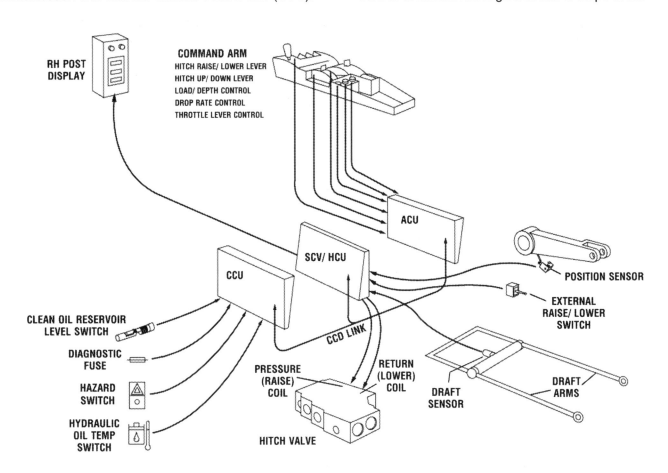

FIGURE 12- 8000 SERIES TRACTOR HITCH CONTROL SYSTEM

determine proper calibrations for all operator control and hitch feedback potentiometers. The calibration steps also determine appropriate electrical deadband for raise and lower actuation of the hitch valve. Once calibrated, the HCU is returned to normal operating mode where it receives signals from operator controls and sensing devices and determines how much current to send to the raise or lower solenoid of the hitch valve.

The hitch valve, mounted at the bottom of the valve stack, is basically a pilot operated 3-way spool valve. It directs flow from the axial piston pump to the hitch cylinders, and from the cylinder to return. A pressure compensator/ inlet check is provided to maintain constant pressure drop on the pressure metering area regardless of hitch loads or pump pressure. A load check poppet is provided to reduce leakage and

subsequent hitch drift. A relief valve limits pressure in the hitch cylinders to protect mechanical components. Two hydraulic pilot valves are provided to shift the main spool. These are proportional pressure reducing/ relieving valves which provide a pilot pressure proportional to current in the solenoid coil. One pilot valve provides pressure to shift the main spool to raise the hitch and the other to lower it.

Three external lift cylinder diameter combinations; (ϕ80mm/ϕ90mm, ϕ90mm/ϕ90mm, and ϕ100mm/ϕ 100mm) are used to provide a range of lift capacities from 45.6kN to 69.6 kN, pending on tractor model and options. Cylinders are double acting with a stroke of 240mm.

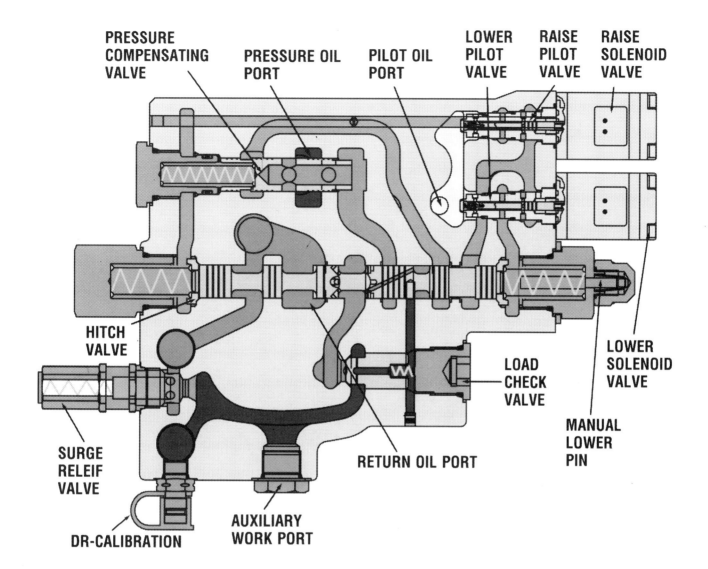

FIGURE 13- 8000 SERIES EH HITCH CONTROL VALVE CROSS-SECTION

SUMMARY:

A new hydraulic system was developed for the John Deere 8000 Series Ag Tractors introduced in August 1994. This system brings together many new custom design components to produce high performance and provide productivity enhancing features.

REFERENCES

Eagles, Derek M.,"Hydraulic System of John Deere 7000 Series Tractors", ASAE Paper No. 931613, 1993.

Reduction of Noise in Hydraulic Systems

Roy G. Wilkes
Wilkes and McLean Ltd.

ABSTRACT

In the past if there was a noise specification to be met, the hydraulic circuit designer selected a pump, from a manufacturer's catalog, that showed a low decibel rating. However as the power unit was manufactured the noise level usually turned out to be higher than that shown in the pump catalog. The problem is that any slight noise generated by the pump is amplified throughout the hydraulic circuit. This pager deals with a new method of stopping the noise before it can be amplified.

INTRODUCTION

A person need only walk through a large automotive company and he will pass some hydraulic power units that have sound enclosures built around them. The sound enclosure is often an indication that the power unit did not operate at the sound rating shown in the catalog for the pump. In other words the catalog pump sound rating is usually much lower than the sound rating for the completed hydraulic power unit.

Something happens during manufacturing a hydraulic power unit that changes a relatively quiet pump into an unacceptable noisy power unit. Pump pressure and pump size have about equal effect on hydraulic noise. However the pump speed has about 300% greater affect on pump noise than either pressure or pump size. This is the reason some pump manufacturers recommend slower electric motor speeds. Fixed displacement pumps are usually quieter than variable displacement pumps. But all of these variables are shown in the pump catalog. In addition to these variables something else is contributing to an increase in noise that is not mentioned in the pump catalogs.

Lab tests show that pump noise levels are increased by 2 to 3 dB(A) just by adding 12 feet of outlet and return lines. The lines do not generate noise, instead they radiate noise when they respond to pulsations or vibrations started by the pump. The pulsations are usually generated by the pump and these vibrations are trans-

ported by the hydraulic lines to large flat metal surfaces were they are converted to the higher noise level.

So not only do hydraulic lines radiate noise but they frequently provide the primary path for propagating noise from the pump to components that, in turn, react to the noise and radiate additional sound. This helps explain why many pump manufacturers have a very low dB(A) pump rating, but when the pump is installed on the power unit the sound rating is much higher.

It is almost impossible to forecast how much additional sound the hydraulic lines and surrounding structures will radiate. This is why many power units are enclosed after they have been manufactured and installed at the customer's plant. The sound enclosure is a result of the additional noise generated by the close proximity of additional large flat machine surfaces.

To get to the source of the majority of noise in any hydraulic system one has to start with the pump. The pump is the main source of the pulsations. The pulsations are the cause of the noise and vibrations. The Piston pump manufacturers seem to have recognized the inherent problem in the piston pump design and are making an effort to correct the noise problem. However after the hydraulic pump designer has done his best, every pump still produces a pump ripple. "Ripple" is the pump manufacturers name for pulsations. It is this ripple or pulsation that produces line vibrations which cause additional noise. Until recently most pump manufacturers ignored the transmission of these pulsations because the methods of trying to eliminate them were too expensive and the transmission and radiation of the pulsations was considered to be the circuit designer's responsibility. This paper deals with new inexpensive methods of reducing the transmission of these pulsations and thus reducing the entire circuit noise level.

One of the first things that should be reviewed when attempting to reduce power unit noise is the hydraulic lines. The one source that contributes the most to adding noise is the incorrect use of hydraulic hose. Recent research at a large pump manufacturer showed that they were able to take an average of 5 dB(A) out of a standard power unit merely by changing the configuration of

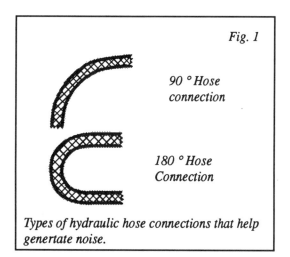

Fig. 1

90 ° Hose connection

180 ° Hose Connection

Types of hydraulic hose connections that help genertate noise.

the hydraulic hose (see fig. 1). In the past, a 90 degree curved hose was used when a horizontal line had to be connected to a vertical line. A 180 degree curved hose was also quite common. Research has shown that both of these configurations actually increase the noise level in the system. The solution is to not bend hydraulic hose but to use a metal tubing for any bend and only use hose

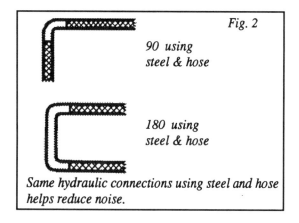

Fig. 2

90 using steel & hose

180 using steel & hose

Same hydraulic connections using steel and hose helps reduce noise.

in a straight line (see fig. 2).

It is common knowledge that introducing a compressible median of nitrogen into the relatively incompressible medium of hydraulic fluid will help reduce pulsations. However the problem here is how do you get the fluid to merely bounce off the nitrogen so the nitrogen compresses and the fluid merely loses its pulsation and nothing more.

Through the years the nitrogen charged hydraulic accumulator has been used in most circuits to absorb hydraulic pulsations. At first the accumulator was used as an appendage device. It hung on the hydraulic line and the designer hoped that the pulsations would enter the accumulator (see Fig. 3). However practice showed that the majority of the pulsations bypassed the line leading to the accumulator. Then various different designs evolved in which the full flow was diverted into the accumulator (see Fig. 4). Sizing this type of accumulator is complicated and the design that diverts the flow into the accumulator is very expensive. The pressure drop through these accumulators is also very high and can create a

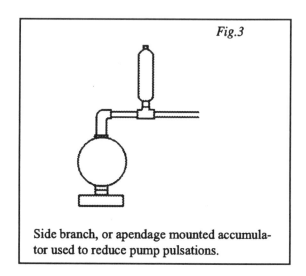

Fig.3

Side branch, or apendage mounted accumulator used to reduce pump pulsations.

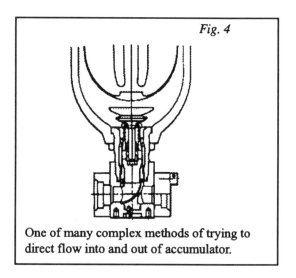

Fig. 4

One of many complex methods of trying to direct flow into and out of accumulator.

problem.

A new method of dealing with the noise causing pulsations is to mount a in-line nitrogen charged noise suppressor right at the outlet of the pump. This new design is a departure from the usual accumulator design. To understand this new design concept, merely think of a hydraulic line with 4158 small radial holes in it and a nitrogen charged bladder pressing down on the holes (see Fig. 5). The holes are large enough to permit fluid flow but too small for the bladder to extrude into them. This new devise is exceptionally small. The length of the unit for a 3/4" size 3000 P.S.I. pipe line is only 8.25". The cost of the unit is usually about one fifth of the price of the larger surge suppressor type accumulators. Another big advantage is that there is no engineering involved in selecting the unit. The size of the unit is determined by the size of the existing hydraulic line. There is a in line suppressor for every pipe and tube size from .375" to 3". In the past the sizing for a hydraulic pulsation dampener was a long and complicated process. With the new design the size of the line becomes the size of the suppressor.

However more important than ease of sizing is how well the unit suppresses the pump pulsations and why it

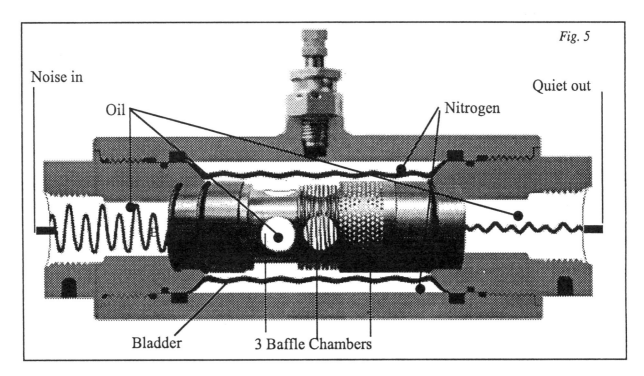

Fig. 5

Noise in

Oil

Nitrogen

Quiet out

Bladder

3 Baffle Chambers

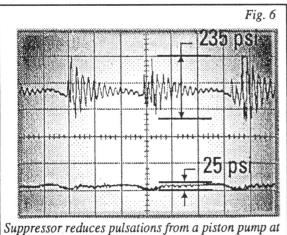

Fig. 6

235 psi

25 psi

Suppressor reduces pulsations from a piston pump at 4,000 PSI at 1800 RPM from 235 PSI to 25 PSI.

works better than other types of accumulators. It is more effective than a much larger accumulator because the fluid only flows .25" radially before it comes in contact with the bladder. Often when using an accumulator the

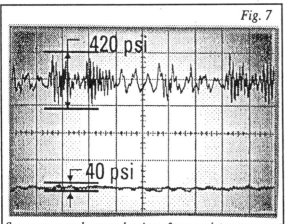

Fig. 7

420 psi

40 psi

Suppressor reduces pulsations from a piston pump at 2,000 PSI 1800 RPM from 420 PSI to 40 PSI

flow path is at least four or five inches long before the fluid contacts the bladder. This short flow path coupled with the fact that there is a much greater flow area leading to the bladder makes the suppressor much more efficient. To date the largest piston pump manufacturers have used the suppressor to help them meet or exceed the strictest automotive noise standards. Figures 6, 7, and 8 show the results of the suppressor tested at pressures from 750 PSI. to 4000 PSI. and at both 1200 RPM and 1800 RPM.

One large pump manufacturer made sixty double pump power units for a large automotive company. When they were completed the noise level registered 82 dB(A). The automotive noise specification that had to be met was 80 dB(A). The pump manufacturer decided that the least expensive solution was to install the new in line

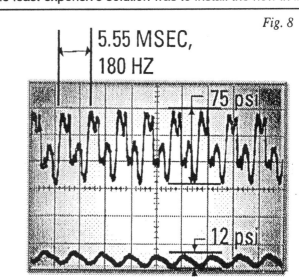

Fig. 8

5.55 MSEC, 180 HZ

75 psi

12 psi

The suppressor reduces pulsations and noise from a piston pump running at 750 PSI at 1200 RPM from 75 PSI to 12 PSI and noise from 85dB to 77dB.

noise suppressor directly at the outlet of the pumps. This simple solution brought the noise level down to 78 dB(A). The cost was considerably lower than building a noise enclosure around the entire power unit.

CONCLUSION

The first thing to consider when trying to reduce noise in a power unit is the configuration of the hoses used in the lines leading from the pump. Whenever possible try to eliminate any curved sections. Replace all curved sections with bent tubing. If additional noise reduction is required consider using a in line noise and shock suppressor. This in line noise suppressor actually does not reduce the noise level of the pump. That is like trying to unring a bell. However it can reduce the pump pulsations that enter the hydraulic system and by reducing the pulsations it reduces the vibrations and thus the noise radiated by the vibrations. So now the power unit designer can look up the noise level of a pump in the manufacturer's catalog and with the use of the suppressor he can expect that will be the total noise level of the hydraulic power unit after it is assembled.